D1195477

Economics and the Design of Small-Farmer Technology

Economics and the Design

of Small-Farmer Technology

Edited by

ALBERTO VALDÉS
GRANT M. SCOBIE
JOHN L. DILLON

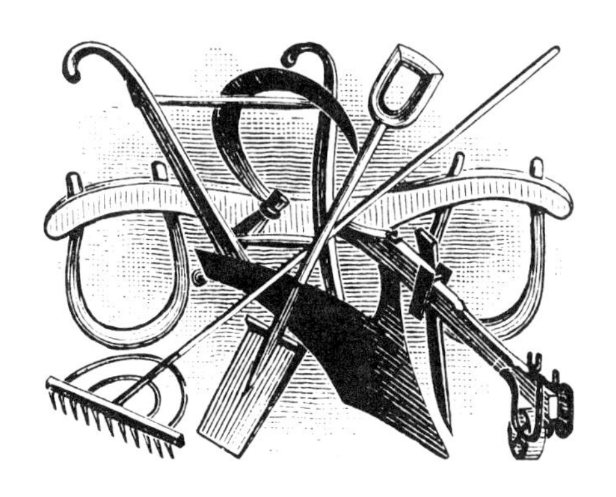

Iowa State University Press, Ames, Iowa

Composed and printed by
The Iowa State University Press
Ames, Iowa 50010.

First edition, 1979

Library of Congress Cataloging in Publication Data

International Conference on Economic Analysis in the Design of New Technology for Small Farmers, Centro Internacional de Agricultura Tropical, 1975

Economics and the design of small-farmer technology.

Bibliography: p.
Includes index.
1. Farms, Small—Congresses. 2. Agricultural innovations—Congresses. 3. Agriculture—Economic aspects—Congresses. 4. Rural development—Congresses. I. Valdés, Alberto, 1935- II. Scobie, Grant McDonald. III. Dillon, John L. IV. Centro Internacional de Agricultura Tropical. V. Title.
HD1476.A3I57 1975 338.1 # 6 78-18917

ISBN 0-8138-1910-5

Contents

Foreword

The "small farmer" or "peasant farmer" is an important client for new technology developed for the purpose of increasing basic food production and improving human welfare in the less developed countries of the world. While statistics may vary depending on the definition of a small farmer, it is clear that the major production of basic food crops in most tropical developing countries is achieved on relatively small farms by people with very limited capital resources. That this type of producer has over the centuries managed to feed self and family, in spite of the vagaries of climate and competition from insects and microorganisms, testifies to an astute understanding of the environment. This has enabled development of complex multiple-cropping systems to provide a consistent supply of food for the farm family in the face of severe hazards and constraints.

Yet small farmers have been much maligned. Their systems of farming are often referred to as "primitive" and they are considered backward, conservative, and lazy because of slowness or reluctance to follow advice of extension agents. We must recognize that, although they may be illiterate, they are not stupid and, although education may be limited, they are often shrewd in knowing what is profitable and what is not. Former U.S. Secretary of Agriculture Orville Freeman once remarked that he had encountered many farmers who could not read—but none who could not count.

Small farmers have often been right in questioning whether technology developed and tested only on experiment stations would work under local operating conditions. They have instinctively realized the unacceptability of taking the large financial and social risks involved when new practices have not been sufficiently tested under existing conditions, when markets are uncertain, and when "cheap food" pricing policies do not provide sufficient incentives. Yet the experience of the Green Revolution in Asia has taught us that when products of appropriate research make new techniques substantially more profitable than traditional methods, and when a remunerative relationship between input and product prices exists, the small farmer will adopt these practices with amazing agility.

It has become increasingly clear that new technology must be so designed that it will take into account the social, physical, and economic realities of the small farmer. The need for increased food production is too urgent and the stakes are too great for physical scientists to develop new technologies without sufficient regard for these realities and then have social scientists conduct ex post analyses regarding their suitability. At CIAT agricultural economists are included as integral components of the multidisciplinary research teams developing new technology for specific commodities. In this way biological scientists and social scientists are working together to develop and validate new technology that will be appropriate under the real conditions of various categories of producers.

Because of the especially difficult task of generation of appropriate technology for small farmers and the key role of economists in this process, a conference on this subject seemed most appropriate. Consequently, a meeting of leading economists involved in this type of activity in various parts of the world was convened at CIAT on November 26-28, 1975. The papers presented at this meeting should have broad interest and utility. It is most gratifying therefore to see that these are now being published in book form. It is our sincere hope that this will represent an important contribution to the urgent task of rural development and improved human nutrition.

John L. Nickel
Director General
Centro Internacional de Agricultura Tropical (CIAT)

Preface

This publication presents the proceedings of an International Conference on Economic Analysis in the Design of New Technology for Small Farmers, held at the Centro Internacional de Agricultura Tropical (CIAT), November 26-28, 1975. The conference brought together 39 participants from 11 countries, and was largely financed by the training and conferences program of CIAT. In addition, valuable support was received from the World Bank, other international agricultural research centers, and the Ford Foundation. The organizing committee, which consisted of the editors, is grateful to the contributors and participants for their cooperation in the conference.

In preparing this publication, the editors have grouped the papers in three broad areas: (1) Methodological Aspects; (2) Design of Technology; and (3) Technology, Rural Development, and Welfare. While, with one exception, the case studies focus on Latin America, it is believed that the analytical approaches presented warrant consideration wherever the problem of design of agricultural technology for small farmers is of concern.

Alberto Valdés
Grant M. Scobie
John L. Dillon

Agency Abbreviations

AID—Agency for International Development
CATIE—Centro de Agricultura Tropical para la Investigacion y Enseñanza, Costa Rica
CIAT—Centro Internacional de Agricultura Tropical, Colombia
CIENES—Centro Interamericano de Enseñanza de Estadística, Chile
CIMMYT—Centro Internacional Mejoramiento de Maiz y Trigo, Mexico
CIP—Centro International de la Papa, Peru
COMECON—Council for Mutual Economic Assistance, Poland
EMBRAPA—Empresa Brasilera Pesquisa Agropecuarios, Brazil
IBRD—International Bank for Reconstruction and Development (World Bank), U.S.A.
ICA—Instituto Colombiano Agropecuario, Colombia
ICRISAT—International Crop Research Institute for the Semiarid Tropics, India
IDRC—International Development Research Center, Canada
IFDC—International Fertilizer Development Center, U.S.A.
OECD—Organization for European Cooperation and Development, France

Conference Participants

Camilo Alvarez P.
CIAT
Colombia

Jock R. Anderson
University of New England
Australia

Tulio Barbosa
Universidade Federal de Viçosa
Brazil

James H. Cock
CIAT
Colombia

Humberto Colmenares
ICA
Colombia

Antonio Dias de Hollanda
Universidade Federal do Ceará
Brazil

Rafael Orlando Diaz
CIAT
Colombia

John L. Dillon
University of New England
Australia
and
Universidade Federal do Ceará
Brazil

David L. Franklin
CIAT
Colombia

Christina H. Gladwin
Stanford University
U.S.A.

Nestor Gutierrez
CIAT
Colombia

Peter B. R. Hazell
IBRD
U.S.A.

Reed Hertford
Ford Foundation
Colombia

Douglas Horton
CIP
Peru

Alain de Janvry
University of California (Berkeley)
U.S.A.

Mario Kaminsky M.
CIENES
Chile

John M. Kirby
Victoria University of Wellington
New Zealand

Rosario Lema N.
Universidad Nacional
Colombia

Diego Londoño
ICA
Colombia

Norha Ruiz de Londoño
CIAT
Colombia

Edgardo R. Moscardi C.
CIMMYT
Mexico

Luis A. Navarro D.
CATIE
Costa Rica

Gerald T. O'Mara
Northwestern University
U.S.A.

Ramiro Orozco L.
ICA
Colombia

Julio A. Penna
Universidade Federal de Viçosa
Brazil

Per Pinstrup-Andersen
IFDC
U.S.A.

Libardo Rivas R.
CIAT
Colombia

James A. Roumasset
University of California (Davis)
U.S.A.

Eugenia M. de Rubinstein
CIAT
Colombia

James G. Ryan
ICRISAT
India

John H. Sanders, Jr.
Universidade Federal do Ceará
Brazil

Pasquale Scandizzo
IBRD
U.S.A.

Grant M. Scobie
CIAT
Colombia

Kenneth Swanberg
ICA/IDRC
Colombia

Helio Tollini T.
EMBRAPA
Brazil

Thomas Twomey
AID
U.S.A.

Alberto Valdés
CIAT
Colombia

Albert Wayne Wymore
University of Arizona
U.S.A.

Carlos A. Zulberti
ICA/IDRC
Colombia

Economics and the Design of Small-Farmer Technology

ALBERTO VALDÉS / GRANT M. SCOBIE
JOHN L. DILLON

1

Introduction

Small farmers constitute by far the majority of agricultural producers in developing countries. Most eke out subsistence on the verge of poverty. The provision of new technology has been widely promulgated as a key component for alleviating this problem. The primary objective of the International Conference on Economic Analysis in the Design of New Technology for Small Farmers was exploration of how economists might contribute.

BACKGROUND TO THE CONFERENCE

Since the early 1960s there has been a growing emphasis on the development and diffusion of agricultural technology in the less developed countries. An expanding network of international research centers has been fostered to complement the role of national research agencies. Partly as a product of this investment, new technologies, especially in wheat and rice, have been diffused. These technologies have had an undisputed impact on total output. Yet several studies have concluded that the ensuing benefits have not been equitably shared among producers. While the technology has apparently been potentially relevant to a wide range of farm sizes, its universal applicability has been limited by subjective, ecological, and institutional constraints confronting the small farmer. While these constraints are readily recognizable in ex post evaluation (at which economists have been relatively successful), their explicit incorporation in ex ante appraisal is a more difficult problem that has received little attention. The conference was organized as one step toward correcting this situation by bringing together a group of economists whose recent research had been directed at various aspects of the problem of technology design. It was felt that such a meeting would recognize the relative merits of alternative approaches to the design of small-farmer technology while simultaneously exposing gaps in the existing concepts and techniques.

From the outset the organizers were conscious of the tremendous breadth and complexity of the issues involved. However, in an attempt to delineate boundaries, the conference was predominantly focused on design at the farm level. At the same time, to counter extreme myopia, an effort was made to consider the role of technological change in the broader setting of rural development and small-farmer welfare.

THE CONFERENCE PAPERS

Small farmers somehow, and apparently without significant difficulty, choose the portfolio of technologies they will use. Certainly these choices are made under differing influences of culture, tradition, and environment in various parts of the world. The mechanisms of choice are doubtless varied also, and the number of technology choices available differs greatly among farmers and regions. All this heterogeneity complicates the task of ex ante technology design and appraisal.

Part I: Methodological Aspects

The aim of ex ante technology design is to formulate new technologies that (1) will be viewed by small farmers as attractive improvements over existing methods and hence be used by them and (2) will be of more positive net benefit to the nation than existing small-farmer technology. In Part I the first of these aims is explored in terms of methodological considerations in applying economic analysis to ex ante technology design and appraisal. The economist's problem is to apply generalized paradigms for establishing the relative merit of alternative choices within a framework that satisfactorily mimics, reflects, or approximates the framework of choice within which small farmers make their decisions. The difficulties of doing this and the possible approaches that might be taken are surveyed by Anderson and Hardaker (Ch. 2). They emphasize the essential complexity of small-farming systems and explore various contexts for the ex ante evaluation of new technologies. Their examination of the array of potential analytical approaches to evaluation leads them to conclude that informed intuition supported by simple modeling in the form of budget calculations is likely to be the most fruitful first step to design and evaluation.

The other two chapters in Part I take a narrower focus and examine the question of risk in relation to small farmers. A number of workers have suggested that risk consideration may be an important influence on small-farmer choice of technology. Using data from the Puebla Project in Mexico, Moscardi (Ch. 3) derives a quantitative measure of the individual farmer's attitude to risk. He shows how this attitude (classified into broad categories of risk neutrality and low and high risk aversion) may be predicted on the basis of socioeconomic characteristics expressed in terms of variables reflecting organizational power and human and nonhuman capital. With this information the need for different technological packages aimed at different farmer groups, in terms of their risk attitude, may be determined. Roumasset (Ch. 4), using data from the Philippines, questions the extent to which variability of returns influences small-farmer choice of technology. He finds that the expected value of returns provides the best predictions of technology choice. However, he emphasizes the importance of providing farmers with information on the variability that may occur, so that they can make adequate judgments about expected returns. Both Moscardi and Roumasset bring out explicitly the need for quantitative understanding of (1) the uncertain environments in which small farmers eke out their existence and (2) the way they react on the basis of their subjective judgments and personal attitudes concerning this uncertainty. Unless these factors are handled satisfactorily, it is most unlikely that ex ante appraisal can adequately reflect small-farmer reasoning on technology choice.

Part II: Design of Technology

The design of technology demands an interaction (sometimes extended over a period excessively long for the preference of those involved) between biologists, economists, and administrators. The economists should participate from the start, the first stage being to identify the target groups (i.e., regions and type of farms) and the resource and institutional constraints the farmers face. In this respect the assumptions about which constraints are to be taken as irremovable become critical. At the second stage this process will allow an initial screening of the potential technologies to be designed in the third interaction stage.

Cock's classical and not so hypothetical fable (Ch. 5) about the production of Bongoyams in Bongoland vividly represents the different phases in the development of a new technology. He shows the nature of interaction between administrators, biologists, and economists and the changing functions of economic analysis in farm technology development. As a biologist who believes economists can help, Cock concludes that economists are not needed in some of these phases. He also illustrates the variety of special skills in economics that are required during the different phases of technology development. These range from farm management to the analysis of world market projections.

Because of the need to handle simultaneously many technologies and input constraints, risk factors, different price levels, and yield variation over time, the authors used computerized procedures in the empirical studies presented at the conference. It should be stressed that in these case studies the question of what technology might best be developed is examined exclusively at the whole-farm level.

In the only chapter dealing with livestock, Valdés and Franklin (Ch. 6) present an ex ante analysis of the impact of introducing improved pastures to the vast tropical savannahs of South America that offer unique settlement possibilities. Computer-based simulation is used to explore the design parameters of cattle technology under varying price and credit situations. The authors conclude that improved pastures deserve attention only as fattening enterprises. In contrast, better husbandry is a less risky source of improvement for the cow-calf operation.

In Chapter 7 Sanders and de Hollanda present an excellent example of the use of linear risk programming to forecast the needed characteristics of a new technology for small farmers in a particular region. Based on the fact that small farmers in Northeast Brazil tend to be located in the worst agricultural areas, they ask biologists to produce a technology with low capital and current expense requirements and a low profit variance. Crop drought resistance is also of paramount importance and suggests the potential of sorghum as a new crop for the region. The authors precisely define the required increase in sorghum yields (albeit very dramatic) necessary to assure adoption and raise the income of the small farmers in the Northeast.

Scandizzo (Ch. 8) is also concerned with Northeast Brazil. Using a mathematical model to analyze the institution of sharecropping, he concludes that it tends to delay economic progress. The model predicts that neither the landlord nor the sharecropper have the incentive to invest in new technology; both may resist technical change. However, it appears that mechanical (laborsaving) innovations are likely to be subject to less resistance than biochemical (landsaving) innovations.

At this stage it is not possible to validate how well these three models may predict actual farmer behavior. But it should be stressed that one of the main assets of the modeling efforts is, not their particular results, but frequently the construction process itself. This forces explicit specification of the different elements of the technology and leads to clearer identification of crucial structural relations. In this context, sensitivity analysis can be a very useful tool for identifying the parameters toward which further biological research should be directed.

Part III: Technology, Rural Development, and Welfare

The apparent importance of aversion to risk by small farmers has been widely stressed. Technologies that satisfy only the criterion of higher net profits may be unacceptable if they involve outlays threatening the subsistence level of small-farmer income. The recognition and quantification of multiple constraints at the farm level then becomes an important phase in the adaptation and validation of new technology. Zulberti, Swanberg, and Zandstra (Ch. 9) draw on their experience in a Colombian rural development project to illustrate these principles. They show that while a proposed corn-production package was supposedly vastly superior in terms of net income per hectare, its adoption by farmers was limited. By analyzing the riskiness of the technology, they hypothesize that the low adoption rate was in part due to a greater expected loss. They conclude by showing how technical and institutional modifications were incorporated to make the technology more acceptable; i.e., by more closely matching the technical and economic requirements of the package with the limitations faced by the small farmer.

The concluding chapters by Dillon (Ch. 10) and de Janvry (Ch. 11) provide a broader perspective of the small-farmer situation. After reviewing the magnitude and structure of the problem and arguing that it transcends economic accounting, Dillon examines three types of theories that bear on the small-farmer problem: dual-economy models, the Schultzian "poor but efficient" view, and the more global theory of unequal exchange. New technology is seen as a driving force in the first two cases and as a possible catalyst for the necessary structural change implied by theories of exploitation. De Janvry examines the theory of unequal exchange in greater detail. He offers a broad historical analysis that makes the social and ecological situation of today's peasantry a legacy of previous exploitation at the national and international levels. On this basis he develops a typology of rural development programs from which he draws implications for the design of technology and institutional change, while examining the role of technology as a vehicle for inducing social change.

UNFINISHED BUSINESS

The conference papers and discussions support the view that economists in cooperation with biologists can make valuable contributions to the ex ante design of new technology for small farmers. Since the 1960s there has been a substantial trend toward more quantitative ex ante modeling of a more formal (though not necessarily more complex) nature, aimed at enhancing technology design. This more formal analytical modeling has been seen as highly advantageous and beneficial because it (1) forces an objective view; (2) leads to consideration of all relevant data; and (3) necessarily demands consistent and logical argument, thereby exposing false analytical assumptions.

A variety of problems were also recognized. Chief among them were questions of research priority, the role of policy, the need for institutional change, and our inadequate knowledge of the resource and psychological attributes of small farmers. These may be posed as a series of open *interrelated* questions needing substantial further research:

1. To what extent, from a global and/or national view, should research priority be given specifically to small-farmer technology? Might not

general welfare gains to consumers (including small farmers) be more readily achieved by concentrating on technology for commercial agriculture?

2. Even if we were successful in developing new technology for small farmers, is there any guarantee that they would capture any of the benefits of this enhanced productivity? To what extent and under what circumstances are changes in agriculture policy needed within the existing institutional structures? Or are changes (perhaps drastic) needed in the institutional structures themselves before improved technology can be of any significant benefit to small farmers?
3. What generality is there in the ways in which small farmers make decisions? How are they influenced by tradition, culture, and socioeconomic circumstance, and what do these factors imply for the design of technology?
4. What is the resource base of small farmers? What are the dynamics of their land resources from an ecological point of view? To what extent are they progressively forced to farm on poorer deteriorating land (as appears to be the case in Latin America), or is their situation more stable (as appears to be the case in Asia)? If small farmers possess limited, poor-quality land resources, does this impose a technical ceiling on the productivity gains that new technology, generated by a given quantity of research, could hope to achieve? Would the total supply of food therefore be smaller if research resources were directed more toward developing small-farmer technologies?

Only with better answers to these questions and how they relate to one another will it be possible to see the problem of small farmers in true perspective and to properly appraise the type of technology needed and its role in alleviating that problem.

I

Methodological Aspects

JOCK R. ANDERSON / J. BRIAN HARDAKER

2

Economic Analysis in Design of New Technologies for Small Farmers

To contemplate the small-farming subsystem of world agriculture means aspiring to a level of generality that is uncomfortable but necessary to open our topic. Still, to quote Weinberg's (1975) Law of Unhappy Peculiarities, "If you never say anything wrong, you never say anything."

SMALL FARMS AS SYSTEMS

The characteristic small farm is tropical, situated in a less developed country (LDC), and operated by a family producing largely for subsistence consumption. Often several crops are grown, multiple cropping and intercropping are widely practiced, and some animals are kept. Excluding the few systems based on monocultures—notably tree crop production—all this adds up to a recognition that each small farm is a complex system. Further, because of the diversity of resource endowments, methods, skills, beliefs, and preferences, small farming in a general or cross-sectional sense is also very complex.

The complexity of small farms has its roots in the number of separate and composite activities undertaken; the number of effective constraints impinging on these activities; the crucial temporal interdependencies among activities; the poor records and information base for decision making; the number of attributes of farm performance that enter the farm family's utility; and last but by no means least, the inevitable lack of certainty in nearly all facets of production, marketing, and life (Mellor, 1969).

Modeling of small farms or theorizing about them thus confronts some fundamental difficulties. The essence of the small farm cannot be captured adequately in simple soluble models. Many equations must be used to specify satisfactory models of small-farm systems, and these may run aground on the

Square Law of Computation (Weinberg, 1975). This states that, unless some simplifications can be made, the amount of computation increases at least as fast as the square of the number of equations, and this in turn may deny solution on even the fanciest of computers. However, the nature of individual small-farm systems (especially farmer risk attitudes) does not permit adequate (simplifying) statistical summary in mere averages through appeal to the Law of Large Numbers.

These arguments, founded on perceived complexity, lead inexorably to the conclusion that small farms are "medium number" systems, subject of course to the Law of Medium Numbers (Weinberg, 1975): "For medium number systems . . . large fluctuations, irregularities, and discrepancy with any theory will occur more or less regularly." This law serves to warn a prospective analyst of small-farm systems that the task will be difficult, may be impossible, and will probably lead to many mistakes.

We return to the intrinsic frustrations of analysis, modeling, and intuitive appraisal of new technologies in the later discussion of alternative models of evaluation. First, we accept without question the premise that in striving for rural development there is a need to design appropriate technologies for small farmers. We do not enter the debate about the extent to which benefits of changed technology may be reaped by people other than small farmers. We now attempt to clarify some ideas on the nature of new technology and the purpose and context of evaluation.

THE NATURE OF NEW TECHNOLOGY

We take "new technology" to be a euphonic expression for a (frequently only slightly) "different way of doing things down on the farm." Many so-called new technologies consist of changed sowing rates and dates, changed rates and forms of fertilizers, etc., so that there is little intrinsically "new" about them. The case for newness is better when genes or machines foreign to the farm are embodied in the changed technology.

Semantics aside, our purpose here is to emphasize that the name of the game really is evaluation of changes in small-farm technology. But what sort of changes? We find it instructive to categorize new technologies in three ways: notional (quarter-baked), preliminary (half-baked), and developed (full-baked).

Notional new technologies are, because of their hypothetical nature, cheap to invent and bounded only by the imagination of the inventor. Since more fully baked technologies usually have their genesis as notions, attention to generating notional new technologies should not be disregarded. Evaluation of this category can range from intuition to analysis, but analytical appraisal is essentially confined to work on models rather than on real systems. Thus work on a formal model may reveal useful insights about a notional change in the physiology of a crop (ensuant perhaps on a notional change in plant architecture), which in turn may aid the orientation of breeding programs and the identification of fruitful avenues of research.

Preliminary new technologies are the unrefined real (as opposed to notional) products of research. Neither testing (perhaps for disease susceptibilities in some target areas) nor evaluation have been adequately completed. Herein lies the main thrust of evaluative work, which is also the thrust of most later discussion here.

Developed new technologies are rare ones that have successfully survived careful and thorough evaluation and await only successful communication to and adoption by the farming community. Once adopted, they are no longer classifiable as new technologies.

EVALUATION

Purpose

Having established that designers and assessors of new technologies necessarily are primarily concerned with evaluating notional and preliminary ones, we should enquire why evaluation is undertaken and for whom. We recognize that ultimately new technology should contribute to national development objectives, but it is also clear that the welfare of the farmers concerned must be considered if they are to be given an incentive to change their ways.

Evaluation of notional technologies is primarily to provide feedback to the people responsible for developing new technologies. When we come to evaluation of the half-baked new technologies, the importance of the small farmer vis-a-vis the research worker must be recognized. We need to establish that a new technology really will improve the farmer's lot. If we are to be successful, we will also need to establish that the farmer can indeed see improvement, i.e., that the anticipated improvements exceed the "just noticeable difference" margin, which may be quite wide in high-risk small farming. The problem of appraising technologies in a social context is examined by Boon (1964), who discusses the use of accounting prices in such evaluations.

Evaluation can be thought of as an on-going process of monitoring seemingly useful changes in technology. It can and should take place in the various contexts discussed below, and it provides information for various groups—most immediately to the developers and purveyors of new technology, then to the various communicators of information (including early-adoption farmers), and with luck, ultimately to the mass of small farmers.

Context

Evaluation of new technologies may be attempted in different contexts, depending partly on who is undertaking the analysis, for what purpose, and whether the technologies are notional or preliminary in nature. A somewhat abstract context of evaluation, removed to some degree from the realities and complexities of the farming environment, will often be especially appropriate for notional technologies. As the technology becomes more fully developed, it becomes increasingly necessary to relate it to the circumstances of the target group of small farmers. We designate the two extreme contexts of analytic evaluation as *in vitro* and *in vivo* analysis, according to whether it is conducted in the glass-walled buildings of research institutes or "down on the farm."

The distinction is to some extent unreal because no worthwhile analysis can be performed without some knowledge of the circumstances of the target farms. On the other hand, the problems of *in vivo* analysis are severe, and it is therefore desirable to avoid confronting these difficulties until they are unavoidable. *In vitro* analysis may allow unsuitable new technologies to be identified and weeded out with minimal analytic effort.

The problems of *in vivo* analysis stem from the diversity of the target group of small farmers. Small-farm systems are characterized by different patterns of resource endowments, production opportunities, skills, beliefs, and preferences. Generalized solutions for such systems are almost impossible to achieve, while the number of farms is generally too large to permit analysis of all individual cases.

Analysts have sought solutions to these problems in various ways, at least three main approaches being distinguishable—case studies, representative farms, and sample surveys.

In the case study approach a few farms are chosen, not so much for their representativeness as for their suitability for analysis. Thus farmers who keep better than average records or are more articulate than most may be selected. The sometimes implicit justification for the case study approach is that, from an intensive study of one or a few cases, insights of general or widespread relevance to the population of farms may be gained. Any unusual features of the particular farm studied are accepted and accounted for in interpreting the results. The value of the case study approach in the present context depends as much on adroitness of interpretation as on skill in analysis.

In the representative farm approach, by contrast, the real or hypothetical farm or farms chosen for study are selected to "represent" the population of farms in some sense (Barnard, 1963; Barnard, 1975; Carter, 1963; Clayton, 1956). Lack of data usually means this representativeness is of a very limited nature, and it remains an open question as to what extent it is possible to generalize from the analysis of such models back to the population. In our view the dangers are considerable, especially when hypothetical farms are used, for then it is all too easy to overlook the impact of the many and diverse "unusual" features of real farms, which nevertheless impinge on the choice of technology.

Finally, sample survey methods may be used to draw a random sample of real farms from the target population (Wheeler, 1950). In principle, the choice-of-technology issue is then investigated for each of these farms. Subject to obvious reservations about the analytical methods used, the results can be related back to the target population, involving well-established statistical rules to assess the confidence that can be placed in the generalizations made. If the sample is based on an appropriate stratification of the target population, this method is somewhat similar to the representative-farm approach, the distinguishing feature usually being that more farms are investigated in the sampling method. This is an important disadvantage. A relatively large sample will often be needed to represent the population of farms with conventionally accepted statistical precision, and the implied analytical load may well exceed the resources available.

The contextual problems of evaluating technology for small farms are severe and rather intractable. Short of nihilism, we suggest that a partial, incomplete, and inadequate analysis may be better than none at all. To quote Weinberg's (1975) Lump Law, "If we want to learn anything, we mustn't try to learn everything."

Alternate Modes

Basically, there are two extreme modes of human problem solving—analytical and intuitive. The analytical extreme has explicit, sequential, and recoverable attributes, while the intuitive extreme has implicit, nonsequential, and nonrecoverable attributes. In contrast with the logical, reductionist, and vertical reasoning of analysis, intuitive thinking relies on holistic impressions, impulsive synthesis, and lateral reasoning (Zeleny, 1975).

As our evaluative aspirations advance from simple, well-structured, static, and deterministic problems toward the complex, fuzzy, dynamic, and stochastic problems epitomized in small farming, the best approach changes from logical, reductionist, and quantitative toward perceptive, simultaneous, and qualitative (i.e., toward an intuitive approach).

Evaluation of changes in complex systems can seldom be tackled within either purely analytic or purely intuitive frameworks. Our intention is to review some points on the spectrum of frameworks for contemplating technological change in small farms. We begin with a review of some methods for partial or *in vitro* analysis, then consider the more ambitious *in vivo* analysis, and finish the review with a discussion of intuitive methods.

IN VITRO ANALYSIS. The "activity analysis" or "programming" model of a farm provides a convenient way of thinking about the problem of selecting appropriate technologies for a given situation. Such a model is defined in terms of a set of available (new or traditional) technologies, each expressed as an activity or process vector. An activity vector is simply an ordered list of the technical coefficients of a defined production technology. For a unit level of the activity it shows the amount of outputs produced and the amount of inputs employed. These activities can be selected and combined in various ways, the set of possible activity mixes (feasible farm plans) being defined by the resource constraints of the farm. In a programming formulation, an optimal mix may be identifiable if an appropriate and amenable objective function can be defined. Choice of a relevant objective function will be discussed below; for the moment we focus on the activity vector concept itself.

An assumption underlying the activity analysis model is that constant returns to scale operate for all activities. Thus different points on a conventional (nonlinear) production function involving different input to output ratios must be represented by different activities. This means that in reality there will be an infinite number of possible activities, but analysis can proceed on the assumption that a representative subset can be selected. Similarly, for completeness, the activity vectors should include all outputs or inputs that have actual or potential value. Moreover, many inputs, like land and labor, are not homogeneous and should be differentiated according to such factors as seasonality or quality. With discretion, an abridged list of inputs and outputs can be selected that facilitates analysis and yet captures the essence of the problem.

Suppose that in relation to a particular farm (or group of farms) we can define (for a riskless world) a set of five possible technologies represented by the five activity vectors in Table 2.1. These activities might correspond to different variety/fertilizer/rotation "packages." What, if anything, can be said about the relative efficiencies of these activities if we know nothing about the resource endowment of the farm in question or about the prices of inputs and

Table 2.1. Activity vectors representing available technologies for a hypothetical farming situation

Farm situation	Activity 1	2	3	4	5
Outputs					
Corn (t)	3	2	3	3	1
Wheat (t)	2	2	2	3	4
Inputs					
Land (ha)	2	2	2	2	2
Labor (days)					
Sowing	3	3	3	3	2
Harvesting	10	10	10	12	10
Animal power (days)					
Sowing	1	1	1	1	1
Harvesting	10	10	10	10	10
Working capital ($)	15	15	16	15	15

outputs? If we assume only that the two outputs have nonnegative values and that the various inputs have nonnegative opportunity costs, it is evident, for example, that activity 1 is more efficient than activity 2 because it has a higher corn yield with the same wheat yield and resource inputs. Similarly, activity 1 dominates activity 3 because, although the yields are identical, activity 3 requires more working capital than activity 1 with no saving in any other resource inputs. Thus activities 2 and 3 are less efficient than activity 1 and can be eliminated from further consideration. On the other hand, no order of efficiency can be established between activities 1 and 4. The latter produces more wheat output than the former but also involves more harvest labor. To assess their relative economy, we would need to know the value of wheat and the opportunity cost of harvest labor. Similarly, no efficiency ranking of activities 1 and 5 or 4 and 5 is possible without more information.

Generalizing, we can say that an activity is efficient if it produces an output and if production of the same quantity of that output by some other activity (or combination of activities) involves either a reduction in the production of at least one other output or an increase in the use of at least one input (or both). Such a definition may allow some poor technologies to be identified, although it tells nothing about the best mix of activities in the efficient subset. Nevertheless, the relevance of the concept of activity efficiency in the evaluation of technology is obvious. Many new technologies are designed to be output increasing. If a higher output can be achieved without a more than proportionate increase in input levels, the new technology will dominate the old in an activity efficiency sense. In applying this test, however, there are some pitfalls for the unwary. It is essential that a relevant list of activity vector components is identified. For example, new dwarf variety A may appear to dominate traditional variety B if we neglect the output of straw. But if A produces less straw than B, and if straw is valuable, say for stock feed or fuel, then no dominance has been established. Similarly, if we had failed to consider harvest labor in Table 2.1, we would have been misled into the belief that activity 4 was more efficient than activity 1.

Although the form of analysis indicated above may permit some less efficient technologies to be identified, it is unlikely to be sufficiently discriminating to order all the technologies of interest. A further sieving requires information on the costs of inputs and/or the values of outputs. Unfortunately, the implied valuation problems are seldom trivial in the context of small-scale farming. We are interested in the economic costs of inputs (measured as their marginal opportunity costs) and the marginal values of outputs. It will generally be reasonable to assume that the marginal value of a normally sold output is equal to the net market price, but it is less clear what value should be employed for subsistence production or intermediate products such as straw. Similarly, for inputs for which a ready market exists, such as fertilizer or chemicals, it will usually but by no means always be safe to assume an opportunity cost very close to the market price. Exceptions arise when the level of use of such inputs by a farmer is constrained by capital rationing or in some other way. However, for flow resources rather than stock resources, marketability is often more problematical. For example, family labor may have no ready market off the farm, except perhaps at busy times when general labor scarcity exists and the marginal value of labor in use on the farm may become very high (Collinson, 1972).

The usual solution to these difficulties is to place prices on inputs and outputs that can be valued with confidence but to leave the remainder as physical entities in the activity vector. Thus for each activity we arrive at a revised vector comprising a net revenue or gross margin coefficient and a reduced list of inputs and outputs that cannot be readily evaluated in money terms. Again the concept of efficiency in activity analysis can be applied, treating the gross margin as an output. Analysis at this stage may permit further inefficient activities to be identified.

By way of illustration we suppose that in the farming region for which the activities of Table 2.1 are defined, corn is worth $50/t and wheat is worth $40/t (irrespective of the volume of production in the target region), harvest labor can be traded at $1/day, and working capital inputs can be costed at their face value. Other inputs are assumed to have less reliably determined opportunity costs. The three already established efficient activities can now be respecified as shown in Table 2.2.

Now it is clear that activity 4 is more efficient than activity 1 since the remaining inputs are identical and the net revenue is higher. We may suspect that activity 4 is also better than 5, but our definition of efficiency allows no ranking of these two technologies because 5 not only yields a lower net revenue than 4 but also requires less sowing labor. To select an optimal mix of activities 4 and 5, we would need to know the availabilities of the inputs in the abridged list. As is well known (see, e.g., Hardaker, 1971), linear programming can be employed to select the appropriate mix for limited resource availability; this and related methods are discussed below.

Before reviewing farm planning methods suitable for technological appraisal, we must reconsider the relevance of the activity analysis framework developed above. Among the many assumptions made, one needs further

TABLE 2.2. Activity vectors respecified, showing activity net revenues

	Activity		
Farm situation	1	4	5
Net revenue	205	243	185
Inputs			
Land (ha)	2	2	2
Sowing labor (days)	3	3	2
Animal power (days)			
Sowing	1	1	1
Harvesting	10	10	10

comment. The analysis thus far has been wholly deterministic in nature. What happens if we try to face up to the reality of a stochastic world? The best we could manage under assumed certainty would be a partial ordering of technologies, and it is still more difficult to say anything definite if risk is taken into account.

First, we must admit that many if not all components of activity vectors are likely to be stochastic in practice. Not only do yields vary, but so do input requirements of labor, animal power, and the like. The general stochastic case presents very considerable problems for the analyst. While progress might be possible, we have elected to confine consideration to the special, if more familiar, case where the main risks can be composed into a stochastic activity net revenue coefficient. Ignoring variations between activities in input requirements, how can risky net revenues be compared? For example, under what circumstances can activity 4 in Table 2.2 be said to dominate activity 1 if both have stochastic rather than deterministic net revenues? Although 4 may have a higher mean net revenue than 1, it may embody the higher risk of a very low or negative net revenue.

The criteria of stochastic dominance (see Anderson, 1974b) are relevant to such a case. By making progressively stronger but still relatively undemanding assumptions about the nature of decision-maker attitudes toward risk, it is possible to derive increasingly powerful rules for ordering risky prospects in terms of their risky efficiency. Alternatively, making the rather strong assumption that risk can be measured by variance of activity net revenues, the more familiar (E, V) or mean-variance criterion can be used to identify the risk-efficient subset of prospects.

It seems possible that there may be some opportunity for combining the criteria of activity efficiency and risk efficiency. For example, given two activities A and B with vector entries comprising stochastic net revenues and resource inputs (some of which are also stochastic), A may presumably be said to be more efficient than B if the distributions of net revenues per unit of each input for A stochastically dominate the equivalent distributions for B in the sense of, say, first-degree (or, given risk aversion, second-degree) stochastic dominance. A special case is when resource inputs are all deterministic and the requirements for A are all less than or equal to those for B with the net revenue of A more risk efficient than that of B. Further developments along these lines might be possible but have not been attempted here, partly because of the entailed difficulties but also because we suspect that any resulting efficiency criteria would not prove at all powerful for ordering technologies.

In vitro analysis provides some means, albeit rather limited, for ordering technologies on efficiency grounds. Unfortunately, it seems likely that often such ordering rules will prove too weak to lead to any worthwhile conclusions. In that event analysts must come down to earth and confront the realities of actual farms and farmers. It will be necessary to take account of the resource endowments of individual farmers, of the alternative uses of these resources, and of their preferences and perhaps also beliefs. In other words, whole-farm decision modeling may have to be attempted.

IN VIVO ANALYSIS. A wide range of whole-farm planning methods exists. Budgeting in one form or another is probably the most widely adaptable technique. Its power tends to be limited only by the capacity of the analyst. Budgeting methods are often thought of as relating to accounting procedures to assess the *profitability* of some change in farm methods or organization, but budgeting also extends to a variety of methods for assessing the *feasibility* of such changes. Thus budgeting can be used to account for limited resource availabilities, interrelationships between activities, etc. The analyst must use judgment to decide which of such relationships will be accounted for explicitly and which will be handled more subjectively or intuitively.

Disadvantages of budgeting may be said to relate to the lack of any formal optimizing algorithm and the difficulty of taking account of risk when using the technique. Because the optimality (or even near optimality) of any budgeted plan cannot be guaranteed, analysis must proceed on a trial-and-error basis. It is possible to extend budgeting methods to incorporate a risk dimension, but the added complexity will generally take the required computations outside the feasible range for noncomputer methods. Models can be adapted for computer analysis in such forms as Monte Carlo budgeting, but only at the cost of some loss of flexibility.

Forms of budgeting have been developed that are designed to bridge the gap between computerized mathematical programming analysis and intuitive planning. The various forms of "simplified programming," also called "program planning," fall into this category. These methods appear to have fallen from favor with increased availability of computers, but we wonder if their potential for use in LDCs, where computers and the skills to use them are less common, has been fully exploited.

In the search for ways to identify the "best" portfolio of technologies for particular farming circumstances, mathematical programming methods have been enthusiastically embraced by academic researchers, if not by practitioners. Linear programming has been the most widely used of these methods. Linear programming provides an intuitively attractive way of solving the activity analysis problem posited earlier. Not only are efficient activities identified, but the mix of efficient activities that uses the available farm resources in an "optimal" way (as measured by a specified linear objective function) is identified. Integer constraints, which often arise in small-scale farming, once limited the applicability of the method, but today operational integer routines that overcome this problem are available.

The principal outstanding criticism leveled against the use of linear programming for whole-farm planning relates to the embodied assumption that all planning coefficients are known constants. In a risky world this may be very far from the truth. Nevertheless, examples of the use of linear programming for farm planning are legion, including applications to planning peasant farms (e.g., Clayton, 1965; Heyer, 1971, 1972; Johnson, 1969; Langham, 1968; Ogunforwora, 1970; Wills, 1972).

In reviewing extensions of the basic linear programming model designed to account for risk, it is useful to differentiate models with stochastic net revenues from models in which risk is recognized in the constraints. The former are generally more straightforward. In the simplest case stochastic net revenues can be handled by changing the objective function to the maximization of expected net revenue, subject to the usual constraints and restrictions. The chief disadvantage of this model, apart from the exclusion of stochastic constraints, is that no account is taken of possible risk aversion by the farmer.

One way that risk aversion can be dealt with when only activity net revenues are stochastic is by use of quadratic risk programming. The procedure is usually applied in a two-stage manner. In the first stage solutions are obtained to the quadratic programming problem (whose objective function is to minimize the variance of the total net revenue), subject to the usual technical constraints and to a minimum expected income constraint varied parametrically over its feasible range. The second stage involves determining the optimal point on the "efficient" set of farm plans so derived. This may be done by direct inspection of the results by the farmer, or the optimum can be located analytically using a Taylor series expansion of the farmer's utility function for income, truncated to account for only the first two moments of the distribution (Anderson et al., 1977).

For the application of quadratic risk programming to a farm-planning problem, a covariance matrix of activity net revenues is required. In most applications (e.g., Camm, 1962; Freund, 1956; McFarquhar, 1961) these data have been obtained from farm records, although Lin et al. (1974) describe a procedure for accounting for the farmer's subjective beliefs about the risks involved. Specification of the required covariance matrix either from past records or by use of subjective probabilities seems likely to be very difficult in most small-farm contexts. Moreover, the usual quadratic risk programming models provide no very appropriate means of accounting for risk in subsistence crops whose output is not converted to cash.

Practical applications of quadratic risk programming have not been numerous, principally because of a dearth of suitable, reliable computer routines; a number of attempts have been made to circumvent the computational problems by using modified linear programming models to account for stochastic activity net revenues. The approaches include the incorporation of game-theory decision criteria into a programming formulation (Hazell, 1970; Low, 1974; McInerney, 1967, 1969), the use of mean absolute deviation in place of variance as a measure of risk (Hazell, 1971), and the use of constraints on maximum admissible loss (Boussard, 1971; Boussard and Petit, 1967). Most of these approaches can be criticized on the grounds that, unlike analyses based on utility maximization, they have no firm axiomatic foundation (Anderson et al., 1977). That is, they are based on computationally tractable formulations of the planning problems, into which the actual problem is forced with arbitrary representation of the farmer's real preferences.

Of the linear programming methods mentioned, the mean absolute deviation approach (MOTAD) escapes the above criticism, since Thomson and Hazell (1972) have shown that the MOTAD programming results approximate those obtained by quadratic risk programming. Schluter (1974) has described the use of MOTAD programming to examine cropping patterns of a group of peasant farmers in India. He claims good agreement between computed and actual farm plans.

When net revenues are not normally distributed and risk-averse utility functions are not approximately quadratic, the quadratic programming and related methods are not really appropriate for planning or for evaluation of changed technologies. Methods to handle such situations have not been well developed. One approach has been to use Monte Carlo programming to generate feasible plans and to review these using the various stochastic dominance rules to dispense with inefficient plans. This approach, called risk-efficient Monte Carlo programming (REMP) (Anderson, 1975), has the feature (which implies both virtues and problems) of producing many diverse plans that are all efficient in the defined sense. Our attempts to use it in evaluating new technologies for small farmers in Northeast Brazil have led us to question its value in such work.

The models discussed so far deal only with risk in the activity net revenue coefficients. Unfortunately, in practice both the technical coefficients and the resource stocks may also be partly stochastic. For example, the labor requirements for weed control in crops may be uncertain, depending on weather conditions, while the time available for this work may also vary according to such factors as the health of family members and other demands on their time. Programming methods to accommodate risky constraints are usually known under the generic name of stochastic programming.

Following Hadley (1964), it is convenient to classify stochastic programming models into sequential and nonsequential problems. Sequential problems involve related decisions made at different points in time, with the property that the later decisions may be influenced by the earlier ones and by the stochastic parameters that only become known after the earlier decisions have been taken. In nonsequential problems all decisions are (or could be) made at the same moment, before the uncertain events occur.

Stochastic programming problems are difficult to solve because of the need to maintain feasibility of the solution vector over a range of random values of technical coefficients and/or resource stocks. Although certain special cases of nonsequential stochastic programming can be solved, the general intractability of this class of problems has led to the development of alternative, approximate formulations. The most common of these is chance-constrained programming (Charnes and Cooper, 1959, 1963), in which the selected objective function is maximized subject to any deterministic constraints specified in probabilistic form such that they are satisfied, either individually or collectively, at some chosen level of probability. Although chance-constrained problems are generally more amenable to solution than other stochastic formulations, the method becomes impracticable if more than a few risky constraints are to be accommodated or if the probability distributions are not tractable (Kirby, 1970). Even simple cases normally require the use of nonlinear programming routines.

The difficulties encountered in nonsequential stochastic programming are magnified in the sequential case; yet by the nature of agricultural production, the problem of farm planning under uncertainty is sequentially stochastic. For example, a farmer's decision on how much of any crop to plant at a given time may depend on his previous planting decisions as well as on such random factors as weather conditions (affecting yield prospects) and perhaps prices over the last few months.

None of the mathematical programming methods currently available is capable of solving the general sequentially stochastic farm-planning problem; although an approximate method, known as discrete stochastic programming, has been developed and applied to agricultural problems (Cocks, 1968; Rae 1971a, 1971b). The method involves specifying a small number of possible states at each stage in the sequential decision process. Decision variables at each stage are linked by appropriate constraints; in the linear programming case, expected total net revenue can be maximized. A nonlinear objective function must be used to account for risk aversion.

The operational value of discrete stochastic programming is limited by the fact that a separate submatrix is needed for each state considered at each stage, with these submatrices arranged in block-diagonal form. Consequently, models quickly reach unmanageable proportions as they run into the "curse of dimensionality." Gross simplifications of reality may have to be made to preserve computability, and even so the resulting matrix may still be very large indeed.

Even when they have been modified to accommodate some recognition of the impact of risk and uncertainty, the above-mentioned budgeting and programming models are very much in the neoclassical paradigm. There has been a definite swing away from this toward more appropriate paradigms that highlight the multidimensional nature of farmer goals and preferences (e.g., de Janvry, 1975; Doyle, 1974; Hazell and Scandizzo, 1973; Lipton, 1968; Sen, 1966).

The dimensions perceived include subsistence and other consumption, leisure, and attitudes toward risk and credit. Adding to the complexity of objective functions in this way (whether explicitly or implicitly) greatly increases the difficulty of modeling in any algorithmic fashion. Not only are the models usually necessarily stochastic and nonlinear, but the dynamic aspects and the consequent combinatorial problems lead to insoluble specifications that are strictly analytical. This should not surprise us—small farms, remember, are medium number systems!

One possible way out of this seeming dilemma is the route offered by the emerging methods of (ad hoc) systems simulation. This is hardly the place to enter upon a review (necessarily lengthy) of this freewheeling approach to "analysis," especially as there apparently has been virtually no simulation work addressed to the evaluation of new technology for small farmers. Simulation methodology has been reviewed by Anderson (1974a) and Rausser and Johnson (1975), while applications in agricultural economics are illustrated in Dent and Anderson (1971) and reviewed in Anderson (1974a) and Johnson and Rausser (1977). In principle, there seem to be no limitations on the methodology of simulation for incorporating the behavioral elements of any pertinent paradigm. Johnson and Rausser (1977) in their review of firm and process simulation models found that about 0.75 were dynamic, 0.6 were stochastic, 0.7 were nonlinear, and 0.9 had a decision orientation. Clearly the approach is very flexible.

However, simulation has some costs that are not always obvious to an intending practitioner. One cost of unconfined nonlinear modeling is the requirement of specifying (estimating?) numerous nonlinear relationships. Similarly, stochastic modeling naturally demands stochastic specification—usually in larger doses than practice has suggested (Anderson, 1976). In short, simulation is no panacea because it seemingly always consumes rather more research resources than anticipated a priori.

One feature that tends to make simulation work expensive and/or tedious is the considerable volume of "results" that are so readily produced. Since this feature is more or less shared by many of the risk-programming approaches (which generally identify a diverse "efficient" frontier or set of plans and technologies), it provides a useful general point with which to conclude this discussion of *in vivo* analytical approaches.

Realistically conceived formal analysis will almost inevitably reveal that many technologies are in some defined sense "efficient." How does the analyst cope with this "result," and how can he use it to distill his own best action on behalf of his diverse audience of small farmers? We wish we knew the answer. Short of throwing up our hands in despair, there seem to be two broad alternatives. First, we can consider carrying forward several "new" technologies as useful developed or recommended ones. This idea of having a portfolio or range of recommended technologies has some appeal in that it recognizes the diversity of farm situations and farmer attitudes. However, extension workers (and perhaps some farmers too) are unlikely to find the idea tolerable, as they will not have a clear-cut "package" to push (or to adopt); if several new varieties are involved, the problems of multiplying and marketing seed will be magnified.

So it seems likely that the second approach of distilling the efficient developed technologies down to just one or, at most, a few readily promoted "new technologies" will generally be preferred. This process seems to us to involve inescapably a highly intuitive or subjective decision. An analytic route thus provides no escape from intuition—an important mode of assessment that we now turn to explicitly, albeit briefly.

INTUITION. At the other end of the spectrum of modes for evaluating changes in technology rests intuition. With intuition, even more transparently so than with analysis, it is most important to know whose is being discussed. The village idiot may be able to apply linear programming methods in tolerable fashion, but it is unlikely that we will be enamored with the holistic impression, impulsive syntheses, and the other stuff and substance of his intuition.

When we entrust our airplanes and lives to the care of the air traffic controller at a busy airport, we are admitting that for this complex problem our faith rests in (skilled) human intuition rather than in an analytic model perhaps implemented on a computer. The controller's experience, fast-functioning mind, and grasp of the simultaneity and totality of the situation cannot be matched by formal models of the system. But this is not to imply that analysis, reductionism, and sophisticated hardware like radars and computers have no place in a situation of human evaluation and control. Very obviously they have a key role in aiding intuition.

To return from the clouds, the sort of evaluative intuition of which we are speaking is founded on intelligence, open-minded perception, pertinent education, experience, and, it is hoped, lots of common sense. It may be supplemented by careful analysis involving some of the models reviewed above. We are claiming that such intuition is underrated, underconfessed, and generally underrecognized as a useful activity. However, what more can be said at this high level of generality?

Apart from the aforementioned general preconditions for good intuition, there seem to be at least two special additional considerations in the context of evaluating new technologies for small farms. First, there is the obvious necessity to know the existing farm scene intimately. This becomes a more difficult and challenging task to the extent that the evaluator does not adopt the stance of a social anthropologist, does not live and work close to existing farming practice, and may not even visit the target community.

Second, and closely related to the first condition, there is the necessity to know the new technology very completely. We can easily comprehend a technologist's enthusiasm for getting hopeful innovations "going" among the audience of small farmers as rapidly as possible. This will probably be before even the technologist has been able to discover the full impact of the new methods in the changed farming environment, perhaps under unfortunate and unanticipated conditions. By the diverse nature of small farming, mistakes are bound to be made, but a good base of knowledge for intuitive evaluation would minimize the chance of error.

CONCLUSIONS

We have sought to show that many analytical and methodological issues in the evaluation of small-farm technologies remain unsolved. Moreover, we have argued that the complexity of small-farm systems implies that these unresolved issues are unlikely to yield readily to attempts to find solutions. Certainly, the methods of mathematical programming, despite recent advances, can at best model small-farm systems only very imperfectly. Nor do more general simulation models provide any panacea, since the costs of such modeling can be very high. Thus the methods of budgeting and other forms of partial analysis have much to recommend them, provided they can approach reality sufficiently to yield insightful answers.

Simple models, whether in the form of budgets or in some other form, have two important advantages. First, they are relatively easy to use and so are not very demanding of special analytical skills, advanced computer facilities, or the like; but they are quite demanding of knowledgeable intuition. This means that these methods can, at least potentially, be used by such people as extension workers, thereby permitting evaluations of technology to be made on a widespread and local scale. Second, simple models permit the chain of causality between assumptions and model output to be more readily traced and understood. Consequently, although the assumptions may be stronger than for more complex models, analysts are less likely to be deceived by their own fabrications. An examination of the results of a model where cause and effect can be related seems more likely to lead to insights into the way the real system works, and such insights will be invaluable in the final and unavoidably intuitive decision-making stage.

COMMENT / *Gerald T. O'Mara*

Anderson and Hardaker have made a very good survey of the several approaches to the analysis of the decision problems of the small farmer in LDCs with respect to choice of earning activities and techniques. However, this fine review is marred by the casual way they examine the economic links of the small farmer with the rest of the economy. This weakness results in a treatment of evaluation of efficiency of new techniques that is deficient and may be misleading. Briefly stated, they neglect the now quite extensive literature on social cost-benefit analysis of investment in developing countries. For a detailed discussion of the two major approaches to project evaluation, the reader is referred to Little and Mirrlees (1969) and Dasgupta et al. (1972) and, for a comparison of the two, to the excellent review of Lal (1974).

Since research, development, and extension activities aimed at developing new technology for small farmers are a form of investment (usually public or at least nonprofit), the need for assessment of expenditures on such activities in terms of social costs and benefits is evident. As Sen (1972) has cogently observed, the evaluation must explicitly recognize the area of control of the evaluator. That is, the accounting prices used in assessing a project must reflect constraints on public policy that cannot be removed. This type of argument leads the United Nations Industrial Development Organization (UNIDO) to recommend that project evaluation assume the continued existence of public policies that are suboptimal, while the Little-Mirrlees recommendations have tended to assume that the evaluator will be able to persuade the government to abandon suboptimal policies. However, the issues are subtler than merely assumptions about future government behavior. Suppose a project would be accepted if continuation of suboptimal policies is assumed (the UNIDO recommendation) but would not be accepted if it is assumed that the government will adopt optimal policies (the Little-Mirrlees position). Then, if the UNIDO position is taken, the project will be accepted and a new vested interest opposed to abandonment of suboptimal policies will have been created. Arguments of this sort clearly suggest that national agencies (and certainly international agencies) concerned with agricultural research and development for LDCs should use world prices as the appropriate accounting prices in evaluating projects. However, this stance implies that substantial obstacles to small-farmer acceptance of new technology will exist in countries with prices that are badly tariff, tax, or subsidy ridden.

The relevance to the Anderson-Hardaker survey of these seemingly abstract comments on project evaluation becomes clearer when it is pointed out that the authors endorse the use of net domestic market prices in the evaluation of a new technique. While these are the prices the farmer faces (if it is permissible to treat the project as an incremental change without effect on market prices), the clear implication (by the authors' arguments) of both approaches to project evaluation in LDCs is that domestic market prices are often poor indicators of social opportunity costs; their use in project appraisal will result in an inefficient resource allocation.

To particularize these points, note that chemical fertilizer input is an almost ubiquitous member of the set of recommended inputs for new techniques promoted to small farmers. Yet owing to the existence of many import-substituting fertilizer projects in LDCs, the price of this input is quite commonly tariff-ridden. Moreover, the prevalence of overvalued domestic currencies and the existence in some countries of fertilizer subsidies, both of which work in the opposite direction, further complicate the fertilizer price picture. In addition, the sharp rise in world fertilizer prices incident to world petroleum prices has resulted in a plethora of quotas, multiple exchange rates, and other ad hoc controls designed to stem the loss of foreign exchange in petroleum-importing nations, while permitting a substantial liberalization of trade controls in petroleum-exporting nations. Under current conditions, the acceptance of the domestic market price of fertilizer as the appropriate accounting price in project evaluation (as applied to a prospective new agricultural technique) would result in an almost random pattern of acceptance or rejection across nations in accord with the erratic variation of domestic fertilizer prices.

The list of badly distorted domestic prices relevant to agricultural project evaluation is almost endless. Many countries subsidize the price of important food grains, while the existence of overvalued currencies, import-substituting tariff structures, and export taxes all tend to depress the domestic prices of agricultural export commodities. For a thorough and well-written review of the manifold effects of suboptimal trade policies, the reader is referred to the study of Little et al. (1970). Domestic factor inputs raise equally troublesome questions. Should labor inputs be valued at market wages, an estimate of farmer's reservations wage, or some planner's estimate of the shadow price of unskilled labor? Where credit is subsidized, should the market price or a shadow price for capital be used in evaluation?

Given the strong arguments for the use of social cost-benefit analysis and its growing acceptance in LDCs, I conclude that economic evaluation of new agricultural techniques must be performed at two levels: the level of the small farmer himself (in financial and/or utility terms) and the level of the government project evaluator (in terms of social costs and benefits). The existence of this two-level filter for the acceptance of new agricultural techniques implies that the problem of an embarrassingly large set of efficient activities discussed by Anderson and Hardaker is at least somewhat (and perhaps significantly) mitigated. It also raises important issues for international centers.

Suppose country A has badly distorted market prices, and evaluators at the international center conclude that the government is unwilling either to remove the suboptimal policy or to make the necessary adjustments in subsidies, controlled prices, or taxes to induce small farmers to adopt a technique that would be optimal if the suboptimal policy were eliminated. Should the center then forget about the small farmers of country A, or should it devise a technique for them based on distorted domestic prices? If it opts for the latter alternative, has it done anything useful for these farmers if a subsequent government should move toward an optimal policy? The answer clearly depends on adjustment costs. For annual crops, these may be small and the latter alternative is to be preferred. For livestock activities or perennial tree crops, the adjustment costs may be substantial and the former alternative may be preferred. The point is that even (and perhaps particularly) international centers must carefully distinguish between domestic market prices and accounting prices (which are dependent on assumptions with respect to future government behavior).

My second major point with respect to the Anderson-Hardaker survey is that it is completely partial-equilibrium in approach. While partial-equilibrium analyses are the rule in agricultural studies, there may be occasions in developing countries when they are not appropriate. If the agricultural sector is large relative to the entire economy and if the new technique will cause a significant resource reallocation if widely adopted (i.e., if it shifts comparative advantage), an economywide or general equilibrium model will be required. More likely is the possibility that the new technique will significantly alter the relative costs and prices of agricultural commodities, violating the notion of an incremental project with negligible effects on the supply and demand of agricultural commodities. Often in this context, a sectoral model using accounting prices for economywide resources (either from a macro model of the entire economy or estimated by using quasi-general equilibrium cost-benefit methods) and embodying estimates of the demand and supply elasticities of the several agricultural commodities will be required, e.g., the Chac model of Duloy and Norton (1973). Or if the effects are localized regionally, a regional submodel may do the job, e.g., the regional spin-offs of the Chac model, *"los hijos de Chac."* All these approaches are encompassed within the class of programming models discussed by Anderson and Hardaker, albeit with some interesting innovations.

A minor point is the neglect of the safety-first approaches to analysis of small-farmer choice under risk. This method can be integrated into the axiomatic expected utility approach under certain reasonable assumptions (see Masson, 1974). Moreover, it is often empirically indistinguishable from an expected utility or a mean-variance approach (Pyle and Turnovsky, 1970), while it can be assimilated into the quadratic-programming techniques (Kataoka, 1963) or even into the linear-programming model using the MOTAD techniques of Hazell (1971).

In conclusion, while it is the role of the discussant to be critical, this duty should not obscure the responsibility to recognize accomplishment. Anderson and Hardaker have given us a remarkably concise and insightful survey of a very broad range of approaches to the analysis of small-farmer decision making and its economic implications, particularly with respect to acceptance of new techniques.

EDGARDO R. MOSCARDI

3

Methodology to Study Attitudes toward Risk: The Puebla Project

The aim of this chapter is to suggest a methodology for studying attitudes of small farmers toward risk. The idea is to be able to evaluate, starting with information provided by a benchmark survey, the extent to which risk may be responsible for differences between the demand for fertilizer without risk and actual demand. Allowance for this effect may then be included in the design of different technological packages.

Risk is introduced in the decision model as a safety-first rule under which an important motivating force of the small-farm family managing its productive resources is the security of generating returns large enough to cover subsistence needs. Under the assumption that the model holds, attitudes toward risk for a sample of farmers are obtained and related to their socioeconomic characteristics. This is an alternative methodology to that by which a von Neumann-Morgenstern utility function is estimated and its parameters related to socioeconomic characteristics. The problem with the utility function approach is that few observations can be obtained, since each is a costly enterprise. The direct method followed here permits the manipulation of numerous ex post observations.

ALLOCATIVE EFFICIENCY UNDER SAFETY FIRST

The model to be used here has been developed by Kataoka (1963), and elsewhere we have given a complete treatment for the case of uncertainty in production (Moscardi, 1975). Under the assumption that the Kataoka-based model adequately describes small-farmer preferences, the value of the risk-aversion measure for the *i*th farmer can be estimated as follows:

$$\eta_{\alpha i} = (\theta/\sigma)_s \cdot [1 - (C_x/P \cdot MP_i^x)] \qquad (3.1)$$

where η_{α} = the risk-aversion measure
θ/σ = the inverse of the coefficient of variation of yields
s = each ecological region
C_x = the price of the xth input
P = the price of the output
MP_i^x = the actual marginal productivity for the ith farmer and the xth input

Since expression (3.1) is obtained as a residual from the behavioral condition of Kataoka's safety-first rule, to the extent that the model does not represent the economic behavior of the farmer, the residual will account for elements other than risk (e.g., lack of information, alternative goals, labor availability, etc.). A *ceteris paribus* condition will in turn be required to take η_{α} as a true risk measure.

To explain the risk-aversion measure, the basic hypothesis here is that risk aversion is a function of the socioeconomic characteristics of the farm family. These characteristics can be represented by endowments of human capital, nonhuman capital, and organizational power. Considering only the monetary value of wealth to explain risk-aversion attitudes is an acceptable procedure for examining portfolio allocation in risky financial markets; however, we think it is an incomplete approach to the study of resource allocation in output markets in the context of small farmers. More specifically, risk aversion is explained as

$$\eta_{\alpha i} = g(D_i) \quad (3.2)$$

where D_i represents the socioeconomic characteristics of the ith farmer.

EMPIRICAL ANALYSIS

Estimation of the Risk-aversion Measure

The source of data for the empirical analysis was the Puebla Project in Mexico (CIMMYT, 1974). The first piece of agronomic information needed to obtain the risk-aversion measure according to expression (3.1) was obtained from the cooperative corn field experiments carried out in zones II and III of the Puebla Project from 1967 to 1971. Zones II and III encompass a quite similar ecological region known as deep soils of the Popocatepetl (see Table 3.1 for rainfall information during the corn-growing cycle). The data covers 1142 observations from 25 experimental trials; each basic observation consists of six values—yield; applied nitrogen and P_2O_5; plant population; and measurements of soil organic matter, pH, and phosphorus content. Table 3.2 contains the means and standard deviations of the observed values.

Table 3.1. Average standard deviation and coefficient of variation of rainfall occurring between April and October for the period 1943–1972 at Huejotzingo, Puebla

Statistic	April	May	June	July	Aug.	Sept.	Oct.	Year total
				(mm)				
$\bar{x}$	*27.18*	*65.85*	*145.33*	*174.49*	*171.03*	*166.01*	*71.55*	*818.80*
σ	21.39	35.66	55.03	70.08	66.03	65.88	51.03	147.40
$\sigma/\bar{x}$	0.79	0.54	0.38	0.40	0.39	0.40	0.71	0.18

Table 3.2. Descriptive statistics from 1142 experimental observations in zones II and III of the Puebla Project

Variable*	Mean	Standard deviation	Coefficient of variation
Yield (Y)	3761.5	1879.4	49.96
Nitrogen (N)	144.4	99.9	69.20
Phosphorus (P_2O_5)	57.5	45.2	78.65
Plant population (D)	50.0	4.33	8.65
Soil pH (pH)	6.63	0.35	5.83
Soil organic matter (OM)	0.54	0.36	65.67
Soil phosphorus (sP)	24.7	12.35	50.01

*Y, N, P_2O_5, and sP are given in kilograms per hectare; D in thousands of plants per hectare; pH in pH units; and OM in percentage.

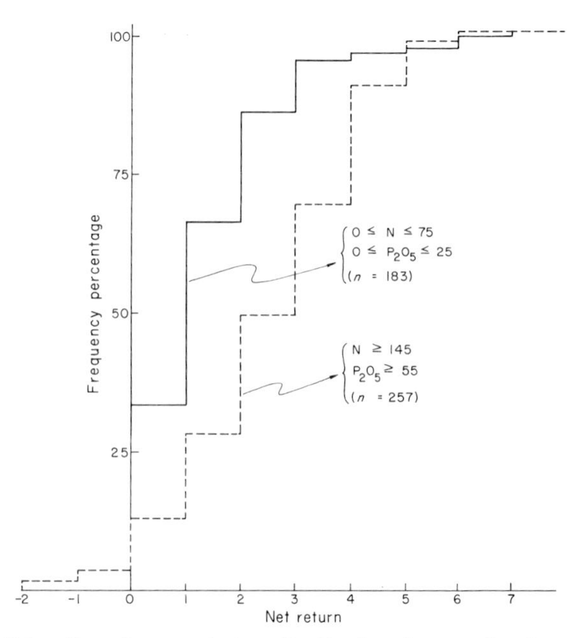

Fig. 3.1. Sample cumulative distribution frequencies for two fertilizer strategies.

To assess the relative riskiness of different rates of fertilizer, two sample cumulative distribution frequencies were calculated for a farmer technology and for a Puebla Project technology. These are illustrated in Figure 3.1. The Puebla Project technology was considered to be those observations with more than 145 kg N/ha and more than 55 kg P_2O_5/ha, and the farmer technology was considered to be those observations with less than 75 kg N/ha and less than 25 kg P_2O_5/ha. As we see from Figure 3.1, the probability of negative net returns is positive for the former technology, while it is zero for the latter.

The basic model employed to estimate the response surface was the Cobb-Douglas model with variable elasticities of production. The production function was required primarily to estimate the marginal productivity for input levels actually used by farmers. The parameters for the production model were finally estimated using ridge regression, due to the problem of multicollinearity. The results are shown in Table 3.3. The next step was to calculate the economic optima for the range of soils included in the experiments to learn the expansion path for the soil conditions of each experimental site. These values are listed in Table 3.4.

The second piece of agronomic information consists of the actual levels of nitrogen and P_2O_5 used by farmers and their location in the project area. This information was provided by a survey conducted in 1971. Forty-five farmers were selected to meet two requirements: (1) those who knew of fertilizer and had applied it in the past and for whom growing corn was the major agricultural activity and (2) those for whom the difference between the available family working time and the family time allocated to generate off-farm income was higher than the labor requirement for growing corn when using the economic optimum levels of fertilizer. By means of this selection we

Table 3.3. Cobb-Douglas model with variable elasticities of production—regular ridge estimates* for $k = 0.2$

Variable	Estimated coefficient
Intercept	0.9778
Log N	0.3423
(OM) Log N	−0.0142
Log P_2O_5	0.0366
(sP) Log P_2O_5	−0.000015
Log pH	1.7098
Log OM	0.0121
Log sP	0.3191
Elasticities†	
$B_1 + B_2(OM)$	0.3346
$B_3 + B_4(sP)^2$	0.0275
$R^2 = 0.60$ $E[L^2_{k=0.2}] = 12.5$ $E[L^2_{k=0}] = 30.0$	

*The ridge estimates for $k = 0.2$ were used for the derivation of economic optimum levels of fertilizer.

†For average values of organic matter and soil phosphorus.

Table 3.4. Optimum levels of applied nutrients by experiments in kilograms per hectare using regular ridge estimates for k = 0.2

Experiment	Nitrogen	Phosphorus
6701	233	19
6704	188	24
6705	191	23
6707	175	23
6708	160	22
6709	177	21
6711	155	22
6712	163	22
6713	190	8
6719	170	3
6725	259	16
6806	150	21
6807	185	24
6808	170	14
6809	198	21
6813	175	19
6927	142	20
6930	151	21
6935	100	12
7004	133	19
7007	172	21
7103	178	15
7104	180	17
7105	182	5
7106	147	20

wanted to keep a *ceteris paribus* condition with respect to information and the opportunity cost of family time.

To obtain the actual marginal productivity of nitrogen for the farmers, each was given the soil condition of the experimental site closest to their location in the project area, and the actual levels of nitrogen and P_2O_5 were adjusted to the expansion path. This measure, the yield coefficient of variation, and the output and input prices of 1971 were used to obtain the risk-aversion measure according to expression (3.1). The results are shown in Table 3.5.

Since Kataoka's safety-first rule implies constrained optima for values of risk aversion greater than zero along the same expansion path, an analysis was conducted to obtain the marginal rates of return associated with diminishing degrees of risk aversion. The results are presented in Table 3.6 and Figure 3.2. The net benefit curve gets flat for expenditures above $600/ha (N = 110 kg/ha and P_2O_5 = 12 kg/ha) and the marginal rates of return drop drastically. It is apparent from Table 3.6 that few farmers will be found to be risk neutral (a risk-aversion measure around 0.20), since the returns for risk taking are very low at this level.

Table 3.5. Economic optimum, farmers' use of fertilizer, corrected rate, and risk measure for five experiments and ten farmers

Farmer	Experiment	Economic optimum		Farmers' use		Corrected*		Risk measure
		N	P	N	P	N	P	
418	6705	191	23	125	100	125	15.0	0.4712
417	6705	191	23	33	20	33	4.0	1.3416
402	6935	100	12	0	0	1	0.1	1.8982
405	6935	100	12	100	15	99	12.0	0.0292
391	7106	147	20	54	40	54	7.0	0.9305
392	7106	147	20	0	0	1	0.1	1.9138
388	6806	150	21	0	0	1	0.1	1.9154
390	6806	150	21	90	0	90	12.0	0.5499
372	6711	155	22	80	40	80	11.0	0.6838
373	6711	155	22	30	20	30	4.0	1.2949

*Corrected at the expansion path.

Table 3.6. Marginal rates of return for diminishing degrees of risk aversion

η_α	N	P_2O_5	Yield	Fertilizer cost	Net benefit	Marginal cost	Marginal net benefit	Rate of return
	(kg/ha)	(kg/ha)	(kg/ha)	($/ha)	($/ha)			(%)
0	190*	21*	4188	1040	1829	109	4	3.7
0.14	170	19	4025	931	1825	108	12	11.0
0.28	150	17	3848	823	1813	113	24	21.3
0.42	130	14	3648	710	1789	108	37	34.3
0.58	110	12	3436	602	1752	109	55	50.5
0.76	90	10	3196	493	1697	109	80	73.4
0.94	70	8	2921	384	1617	109	119	109.2
1.14	50	6	2589	275	1498	112	194	173.2
1.38	30	3	2141	163	1304	109	373	342.2
1.70	10	1	1439	54	931	54	475	879.6
1.99	0	0	667	0	456	...	...	...

*Unconstrained optimum.

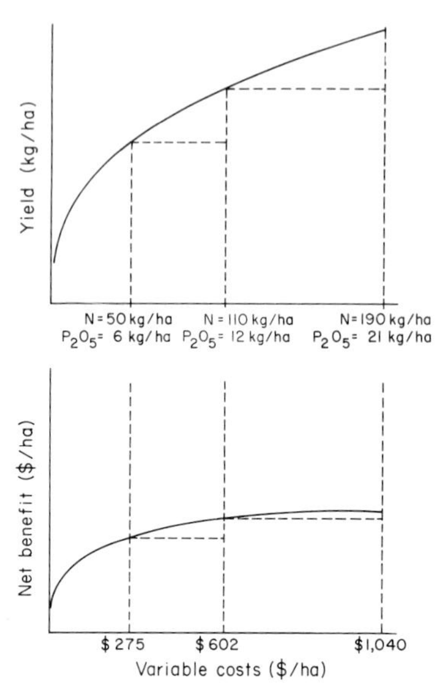

Fig. 3.2. Yield curve and net benefit curve for different fertilizer costs on an average soil.

Discriminant Analysis of Socioeconomic Variables

The basic premise of this chapter is that risk aversion is a function of the socioeconomic and structural characteristics of the farm family economy as indicated by expression (3.2). Three classes of explanatory variables are considered: human capital, nonhuman capital, and organizational power. Under the first group we include age, schooling, and family size.

It is generally hypothesized that, other factors being constant, older farmers tend to be less prone to accept risk than younger farmers. In the context of this study, age cannot be taken as an indicator of on-the-job experience, which may be thought to be positively associated with risk bearing, since ability to farm in the conditions under study does not require a lot of experience and opportunities to develop other types of skills are not easily available.

Higher levels of education have generally been associated positively with risk bearing. In the area under study the average extent of schooling was 2.4 years. A priori knowledge that the older the farmer, the lower the education, was supported by data that showed a partial correlation coefficient between age and education of −0.34.

Family size is a variable for which two interpretations can be given. One is that the greater the size of the family, the lower the willingness of the farmer to accept risks. This would be the case when family size mainly reflects the consumption needs of children. Another interpretation would be valid when family size reflects the labor capacity of the farm family; under this situation a greater family size would imply a higher capacity to generate off-farm income and a higher availability of labor, particularly at harvest time when there might be a shortage of labor in the region. Thus, under this second interpretation, the greater the family size, the higher the willingness of the farmer to accept risks. The analysis of the data supports the second interpretation. On the one hand, the average family size was 5.4 members, with an average of only 1.6 members below 10 years of age. On the other hand, the partial correlation coefficient between family size and off-farm income was positive and equal to 0.25.

Under the group of variables defining nonhuman capital we have total land under control and off-farm income. Total land under control is the amount cultivated by the farm family independent of the type of land tenure. According to the 1967 survey, almost 40 percent of the farmers had a combined private and *ejido* holding, 30 percent were *ejidatarios* only, and 30 percent were private smallholders only. The average amount of land under control for the sample under study was 2.9 ha, including 0.30 ha under irrigation. The hypothesis here was that the larger the area of land under control, the higher the capacity of the farmer to bear risks. The reasoning is that more land permits better chances to spread risks, either because of the possibility of cultivating more or the same amount of one crop under different technologies or because more land implies different parcels at various locations with different kinds of soil and perhaps climatic conditions.

Off-farm income is a variable assumed to have an important impact on risk bearing. According to the sample under study, 65 percent of the farmers generated off-farm income in 1970. This includes off-farm wage income plus other nonfarm income, mainly from domestic, commercial, and industrial employment in nearby cities. Information provided by the 1967 survey showed that 40 percent of the total family income was derived from off-farm activities. This part-time agricultural system is likely to be found in agricultural areas with two characteristics: low quality of the climate-soil system (relatively high agronomic risk) and a relatively high family-size/farm-size ratio. Under this situation the farm family can allocate its working time between farming activities and labor-market activities. The opportunities for income generation outside the family-farm agricultural activities can be in the agricultural sector or in the urban sector. The hypothesis is that the higher the off-farm income, the higher the willingness of farmers to accept risks.

The data showed the following partial correlation coefficients between off-farm income and the other variables included in the analysis: with age −0.20, with schooling 0.23, with family size 0.25, and with total land under control 0.75. The correlation with schooling, although positive, is not as high as one would think, and the positive and high correlation between off-farm income and land under control contradicts all our presumptions. The explanation for these apparently abnormal facts is to be found in the structural characteristics of the area under study. The observed off-farm jobs do not require special training and, given the *minifundia* structure, even the larger farms cannot absorb the available family labor. The relatively high correlation between off-farm income and land under control may be due to the fact that, even among smallholders, more land implies a larger set of opportunities to find some type of off-farm job.

Only one variable is included in this study to account for organizational power: participation in a solidarity group. This comes from the fact that one of the components of the Puebla Project strategy is to provide organizational help for the farmers so that they can obtain the necessary agricultural credit to buy fertilizer. Solidarity groups are formed to this end, relating the farmers with the credit institutions under the supervision of a member of the Puebla Project team. Before making any hypothesis on this variable, a word on its relation to risk aversion is in order. In this study we are using data from 1970, and in that year we had almost 30 percent of the farmers in solidarity groups in the area being studied. We do not know whether the farmers joined the groups because they wished to apply the Puebla Project recommendations or because they wished to have easier access to credit and were already using relatively high applications of fertilizer. If the latter case applies, there would be no relation between risk bearing and solidarity groups. If the former is the case, we hypothesized that a farmer within a solidarity group would be more willing to accept risks. The reasoning behind this hypothesis is that through organizational processes farmers acquire valuable experience that could enlarge the feasible set of opportunities, making it easier to face uncertain situations.

The mean and variance of the socioeconomic variables included in the analyses that follow are presented in Table 3.7. To explore quantitatively the relation between risk aversion and the socioeconomic characteristics defined above, a discriminant analysis was conducted. This technique uses a classification of the data in various groups to determine a function that will

Table 3.7. Mean and standard deviation of the socioeconomic variables included in the analysis

Variable	Mean	Standard deviation
Risk aversion (units)	1.10	0.64
Age (yr)	49.6	14.8
Schooling (yr)	2.4	1.9
Family size (members)	5.4	2.7
Off-farm income ($Mex)	2425.0	5679.03
Land under control (ha)	2.9	3.8
Solidarity group (%)	0.30	...

Note: Sample size was 45.

give the probability of membership in each group for given explanatory variables. In this study the risk-aversion measure has been used as a classification variable to discriminate among three groups: a risk-neutral group, a low-risk-averting group, and a high-risk-averting group. The groups are defined by three intervals for the risk-aversion measure as follows: $0 < \eta_\alpha < 0.40$; $0.40 \leq \eta_\alpha \leq 1.20$; $1.20 < \eta_\alpha < 2.00$.

The general hypothesis that risk bearing is associated with the farm family's socioeconomic characteristics, as well as the particular hypotheses on each variable, will be supported to the extent that the developed discriminant function assigns the observations to the same groups as the classification variable does.

Table 3.8 shows simple statistics for the three groups defined. Prior probability of membership and sample size for each group are given. The grouping strategy is based on the idea that for each farmer there is a range of variation of the risk measure. This variation may be due either to different price conditions and soil characteristics or to factors outside the human nature of the individual, but for a given endowment of human and nonhuman capital and organizational power. A close examination of Table 3.8 shows how the means of the different variables change in the expected direction among groups. Particularly notable are the differences in off-farm income, land under control, and membership in a solidarity group.

Table 3.9 shows the posterior probability of membership in each group for the 45 farmers included in the analysis. It also shows that the explanatory variables considered discriminate between the groups rather well, judging from the values of the probabilities.

Table 3.10 presents a summary of the classification performance of the discriminant analysis. As we can see, the performance is quite good. Of the initial observations assigned to each group, 83 percent remained for the neutral risk group of 6, 86 percent for the low-risk-averting group of 21, and 82 percent for the high-risk-averting group of 18. No change occurred between the extreme groups.

CONCLUSIONS

Estimation of risk aversion shows that this variable is highly responsible for differences between the demand for fertilizer without risk and actual demand. The low percentage of farmers found in the risk-neutral group is seen as evidence for the argument that as economic optimum levels of fertilizer are approached the payoff for risk taking is very small.

Socioeconomic and structural characteristics are important variables explaining the risk-bearing capacity of farmers. The levels of human and

Table 3.8. Discriminant analysis: Mean and standard deviation of the socioeconomic variables for three groups

	Neutral risk group ($Pr = 0.13$, $n = 6$)*		Low-risk-averting group ($Pr = 0.46$, $n = 21$)		High-risk-averting group ($Pr = 0.40$, $n = 18$)	
Variable	Mean	Standard deviation	Mean	Standard deviation	Mean	Standard deviation
Age (yr)	42.8	10.0	47.7	13.6	58.2	12.9
Schooling (yr)	3.2	2.1	2.8	1.7	1.4	1.4
Family size (members)	5.0	2.8	6.4	2.5	4.9	3.2
Off-farm income ($Mex.)	3218.3	3603.5	2669.6	7459.2	1388.7	3586.6
Land under control (ha)	3.91	1.9	2.92	3.9	1.79	1.6
Solidarity group (%)	0.66	...	0.14	...	0.11	...

*Pr denotes prior probability of membership corresponding to frequency in sample denoted by n.

Table 3.9. Discriminant analysis: Posterior probability of membership in each group

Farmer	From group	Classified into group	Group 1	Group 2	Group 3
346	1	1	0.8867	0.1109	0.0025
351	2	2	0.1243	0.8757	0.0001
352	2	2	0.0018	0.8974	0.1009
353	2	2	0.0362	0.8836	0.0802
354	2	2	0.0000	0.9984	0.0016
356	2	2	0.0000	0.9393	0.0607
358	3	3	0.0000	0.1682	0.8318
359	1	1	0.9960	0.0005	0.0035
360	2	3	0.0000	0.1157	0.8843
361	2	2	0.0000	0.5840	0.4160
365	3	3	0.0000	0.2362	0.7638
370	2	3	0.0000	0.1956	0.8044
372	2	3	0.0000	0.0639	0.9361
373	3	3	0.0000	0.0101	0.9899
375	2	2	0.0000	0.6972	0.3028
377	3	3	0.0000	0.0000	1.0000
378	2	2	0.0000	1.0000	0.0000
380	3	3	0.0000	0.1057	0.8943
387	3	3	0.0000	0.0138	0.9812
388	3	3	0.0000	0.4401	0.5599
390	2	2	0.0000	0.7218	0.2782
391	2	2	0.0000	0.9258	0.0742
392	3	3	0.0000	0.1704	0.8296
393	2	2	0.0000	0.9371	0.0629
394	2	3	0.0000	0.1378	0.8622
396	2	3	0.0000	0.0152	0.9848
402	3	3	0.0000	0.2073	0.7927
404	2	2	0.0000	0.9931	0.0069
405	1	1	0.9719	0.0279	0.0002
406	1	1	0.9999	0.0000	0.0000
408	3	3	0.0000	0.0111	0.9889
409	3	2	0.3229	0.5866	0.0905
411	3	3	0.0000	0.0615	0.9385
413	2	2	0.0000	0.9991	0.0009
414	3	3	0.0027	0.0552	0.9422
415	2	3	0.0000	0.2099	0.7901
417	3	3	0.0000	0.0077	0.9923
418	2	2	0.0001	0.5504	0.4495
419	3	2	0.0144	0.7907	0.1949
420	2	2	0.0000	0.9594	0.0406
421	3	3	0.0000	0.0120	0.9880
422	3	3	0.0017	0.0029	0.9953
423	3	3	0.0000	0.3336	0.6664
425	1	2	0.4440	0.5546	0.0014
427	1	1	0.9984	0.0012	0.0004

nonhuman capital and organizational power associated with the farmers of the risk-neutral group (as compared with levels for farmers of the high-risk-averting group) is seen as supporting the above statement.

Some of the findings of this study may be used in the design of Puebla-type projects. According to this research, a classification of farmers in rela-

Table 3.10. Discriminant analysis: Summary of classification performance using generalized square distance

From group	Classified into group 1	2	3	Total*
1	5	1	0	6
2	0	15	6	21
3	0	2	16	18
Total†	5	18	22	45

*Total observations in each group assigned by the classification variable.
†Total observations in each group assigned by the developed discriminant function.

tion to their different attitudes toward risk can be made empirically and used either to design technological packages with various levels of risk or to evaluate the chances of success of a technological package whose adoption requires risk-neutral behavior. Provided basic information from a benchmark survey is available, the parameters of the discriminant function of this research can be used in a similar area to identify different groups of farmers and to assess their relative importance, so as to decide on the need for separate technological packages.

COMMENT / *Christina H. Gladwin*

This is an interesting and imaginative study in the decision-making field, where expected utility has distracted people for years. Moscardi's test of Kataoka's safety-first rule is a welcome change.

However, I have three questions about this study:

1. Does Moscardi actually test the hypothesis he sets out to test, i.e., that Kataoka's safety-first (SF) rule is the way to quantify risk or to formulate a risk constraint?
2. Do his results support his hypothesis?
3. Even if evaluation of the Puebla Project is not his aim, shouldn't he test another hypothesis, i.e., about how important risk is as a limiting factor to adoption of new technologies? After all, the difference between adoption and nonadoption is not risk alone.

Moscardi could have tested Kataoka's SF rule in two ways.

$$\begin{aligned}&\text{maximize} \quad d_i, \text{ a minimum level or subsistence level of income}\\&\text{subject to} \quad Pr(y < d_i) < \alpha_i, \text{ some probability of ruin}\end{aligned} \tag{3.3}$$

First, he could have found a way to empirically determine α_i and d_i for farmer i and used the rule to predict adoption behavior or decisions. Or second, he could have substituted more easily measurable variables for α_i and d_i to test the rule (which is what Nerlove's partial adjustment and revised expectations hypotheses are all about: he substituted more easily measurable variables than expected prices to test a hypothesis about them).

Moscardi (1975) has taken the second route by using the Chebychev inequality "in a sort of free-distribution approach to find a certainty equivalent for the probability of disaster." This approach is too free for me. What Moscardi actually does is turn a test of a SF rule into a test of a mean profits–standard deviation of profits analysis, i.e., a version of mean-variance analysis. Thus Kataoka's SF rule becomes: maximize a preference function $V(\mu,\sigma_\pi) = \mu - \eta(\alpha)\sigma_\pi$, where π is a profit random variable. I do not object to this test of the preference function or of $\eta(\alpha)$ as a risk measure, but I would like to see the derivation of what Moscardi calls the Chebychev upper bound:

$$Pr(\pi \leq d) \leq \sigma_\pi^2/(\mu - d)^2 \leq \alpha \tag{3.4}$$

from the Chebychev inequality, which says

$$Pr[X - E(X) \geq c] \leq \mathrm{Var}(X)/c^2 \tag{3.5}$$

where $c > 0$. Without his assuming either that $\mu_y = \mu_{\pi\text{ corn}}$ or that $d_y = \mu_{\pi\text{ corn}}$, to get from equation (3.5) to (3.4), I think he substitutes $\mu_y - d_y$ for c, a reasonable substitution saying that mean family income is greater than a subsistence level of income by some positive number c. But if he also assumes either that mean profits from corn equals mean family income or that a subsistence level of income equals mean profits from corn to get from equation (3.3) to (3.2), then I question the validity of these assumptions and hence the derivation.

There is a cognitive and, in the Puebla area, a numerical difference between mean income μ_y, mean profits from corn $\mu_{\pi\text{ corn}}$, and a subsistence level of income d_y. One can easily see on a village level in Puebla that some farmers are saved from falling below a subsistence level of income in a bad corn year by their income from milk, pigs, and off-farm income. So one can assume that:

$$d_{y\text{ farmer }i} = f(\mu_{y\text{ farmer }i}) \tag{3.6}$$

$$\mu_{y\text{ farmer }i} = f(\mu_{\pi\text{ corn + milk + animals}_i},\ \text{off-farm } y_{\text{farmer }i}) \tag{3.7}$$

Equation (3.6) says that the subsistence level of income of farmer i is some function of mean income of farmer i; equation (3.7) says that mean income of farmer i is some function of mean profits from corn, milk, and animals of farmer i and his off-farm income. As can be seen in equation (3.7),

Table 3.11. Percentage contribution of four components to the total farm family income in Plan Puebla

		1970	
Component	1967	All farmers	Farmers on credit lists
Net income from crops	30.4	35.5	51.8
Net income from animals	28.4	30.0	16.1
Off-farm income	40.7	27.7	27.1
Miscellaneous income	0.5	6.8	5.0

Source: CIMMYT (1974).

a substitution of μ_π for μ_y, and thus for d_y, ignores mean profits from milk and animals, which was 30 percent of the average family income in Puebla in 1970 (see Table 3.11), and, more important, off-farm income, which was 27.7 percent of average total family income. Table 3.11 shows that income from crops was only 35.5 percent of average total family income in 1970, an increase of 5.1 percentage points over 1967. So an assumption that either $\mu_{y_i} = \mu_{\pi_i}$ (mean income = mean profits) or $d_{y_i} = \mu_{\pi_i}$ (subsistence level of income) is a simple function of mean profits from corn is not supported by the data from Puebla. Hence I question Moscardi's use of the Chebychev inequality to transform Kataoka's SF rule into the preference function that he maximizes.

By assuming that $d_y = \mu_\pi - K\sigma_\pi$, where π is a normally distributed random variable, one can also arrive at Moscardi's preference function $V(\mu_\pi, \sigma_\pi)$. But this assumption *itself* changes the whole meaning of the behavioral assumption from one that says that farmers perceive the risk of adopting new technology as the probability that they might fall below some subsistence level of income into an assumption that says that farmers perceive the risk of adopting new technology as an increase in the spread or variation of profits from corn around the mean, and so will trade off mean profits and risk, i.e., mean-variance analysis. Whatever one may think of the relative merits of these two behavioral assumptions, one should realize they are different. Hence I feel that, although the risk measure η_α that Moscardi uses is a valid measure of risk aversion, α is not a measure of the probability of ruin in Kataoka's SF rule.

Do Moscardi's results support his hypothesis? He gets a measure of risk aversion but does not test how well this measure can explain adoption behavior. Instead, he gets the η_α measure from nitrogen fertilizer production data and tries to explain *it* by regressing it on individual farmer characteristics: $\eta_{\alpha i} = f$(human capital$_i$, nonhuman capital$_i$, organizational power$_i$). Because of the limited range of the dependent variable, he does not try to explain $\eta_{\alpha i}$ via ordinary least squares. Instead he uses a discriminant analysis, in which only 3 out of 6 independent variables are significant at the 5 percent level, that predicts via these variables into which of three groups farmer i will go. I do not know what putting people into three groups tells me.

While Moscardi defines the groups as being a risk-neutral group, a low-risk-averting group, and a high-risk-averting group, they could just as well be defined as a group with high capital availability, a group with medium capital availability, and a group with low capital availability.

Finally, I question Moscardi's methodology because he does not use his risk constraint to ask how important risk is as a limiting factor to adoption. The difference between adoption and nonadoption is not risk alone; yet Moscardi's analysis implies, but does not test, his implication that risk aversion explains nonadoption of high levels of fertilizer in the production of corn. Table 3.12 (from my sample in one village in Puebla, given my three models of adoption-decision behavior) shows that nonprofitability is the important limiting factor in two out of three decisions to adopt the recommendations of Plan Puebla. In decision 1 (i.e., to increase fertilizer use) and decision 3 (i.e., to fertilize twice) nonprofitability of the recommendation is a much more important limiting factor to adoption than risk aversion. Only in decision 2 (i.e., to increase plant population) is risk as important a limiting factor as nonprofitability of the recommendation.

To answer the question of how much risk improves our ability to explain farmer behavior relative to models that ignore risk, we can see from Table 3.12 that including a risk constraint in these decision models improves the predictability of the decision to increase fertilizer use in the future by 22 percent (12 + 10), the decision to increase plant population by 8 percent, and the decision to increase the number of fertilizer applications by 21 percent (6 + 15). A 10 to 25 percent improvement in predictability makes inclusion of a risk constraint in a decision model imperative, since one wants to be able to explain 90 percent or more of the actual decisions made by a group of decision makers.

Table 3.12 also shows that adoption is innovation specific. One cannot see how important a limiting factor risk is by lumping all innovations together and studying adoption of only one, which is to some extent what Roumasset (Ch. 4) and Moscardi both have done by confining their analyses to nitrogen fertilizer use. So we must look at a specific innovation to determine how important a limiting factor to adoption risk aversion will be.

We can only see how important risk is to adoption by formulating a risk constraint and using it (with other factors, e.g., profit, availability of capital or credit, etc.) to build a model to predict adoption or nonadoption by farmers. If we can predict that 90 percent of a group of farmers will or will not adopt an innovation, using a model that includes the risk constraint we want to test (like Kataoka's SF rule), we have a very convincing test of a risk measure or constraint. Moscardi, however, used adoption data in a model to derive a risk measure, tried to explain the measure, and stopped. If he had then used data on the individual characteristics of a second set of farmers to predict their amount of risk aversion $\eta_{\alpha i}$, used this estimate to predict level of

Table 3.12. Relative importance of profitability, risk aversion, knowledge, and capital availability in adoption-decision behavior

		Number of tryers who:		Number of tryers who said recommendation X is:				
Decision	Number of test cases	Pass profit, risk knowledge, and capital constraints	Say recommendation X is risky but try it anyway	Not profitable	Too risky for plants	Too risky to lose cost	Too costly (lack of capital or credit)	Fail the knowledge requirements of recommendation
1. To increase fertilizer use								
In the past*	77	27	5	7	2	0	20	0
		35%	6%	9%	3%	0%	26%	0%
In the future	41	14	2	16	5	4	credit is	0
		34%	5%	39%	12%	10%	assured	0%
							available	
2. To increase plant population	26†	3	1	4	2	0	1	14
		12%	4%	15%	8%	0%	4%	54%
3. To increase number of fertilizer applications	34	8	0	18	2	5	1	0%
		24%	0%	53%	6%	15%	3%	0%
4. To adopt after trying decisions 2 and 3	11	2††	0	7	1	0	0	0
		18%	0%	64%	9%	0%	0%	0%

*The limiting factor is unknown in 6 cases; farmers were still testing the increase in 6 cases; a smaller q_f/ha in 3 cases. One farmer said an increase was not profitable but tried it anyway.

†The limiting factor is unknown in 1 case.

††Both adopted planting closer together, decision 2. One farmer is still testing.

fertilizer use, and then compared their predicted levels with actual levels of fertilizer, we could have seen if his model of risk aversion predicted level of fertilizer use. Lacking this test, there is no way we can know if Moscardi's preference function can be used in a model to predict whether farmers in Puebla will adopt any or all of the recommendations of Plan Puebla.

JAMES A. ROUMASSET

4

Unimportance of Risk for Technology Design and Agricultural Development Policy

The purpose of this chapter is to assess the importance of including measures of risk and risk aversion in descriptive analyses of farmer behavior. Such an assessment requires answers to two subsidiary questions. First, how much does risk improve our ability to explain farmer behavior relative to models that ignore risk? Second, if risk does improve the explanatory power of models, how much difference does it make for formulating agricultural policy? This last question is essential whether or not risk induces correctable market failure.

WHAT IS RISK?

There are two prevalent and equally correct definitions of risk. In high theory, risk is what increases when the density function of returns is subjected to a mean-preserving spread (Diamond and Stiglitz, 1974). For some special cases this type of risk is equivalent to variance. In insurance parlance and in common usage, risk is the probability that returns will fall below a specific level, e.g., below zero or some subsistence requirement.

Modern decision theory does not distinguish between objective and subjective probability. Probability is simply the degree of belief in a particular outcome; i.e., it is always a personalized concept (de Finetti, 1968). The distinction between decision making under risk and decision making under uncertainty, commonly (but falsely) attributed to Knight (1921), also is no longer regarded as useful. Uncertainty refers to the state of mind of a decision maker who perceives more than one possible consequence of a particular act. It is represented in decision theory as a probability distribution. Since risk is a parameter of the probability distribution (e.g., variance, probability of loss,

I am indebted to William Moss for helpful comments.

etc.), risk is likewise a property of uncertainty. Thus the term "decision making under risk" is somewhat a non sequitur.

On the other hand, it is useful to distinguish between situations where the decision maker has had substantial experience regarding the various outcomes and situations where he has very limited information on which to base his subjective probabilities. Thus just as risk can be regarded as a characteristic of uncertainty, knowledge (i.e., how much the decision maker has learned about the likelihood of various outcomes) is a second variable by which uncertainty can be characterized.

Few economists would disagree with the contention that representing a decision problem under uncertainty with a complete certainty model is likely to be misleading. (If many outcomes are possible, which is relevant to insert into the full-certainy model?) It does not follow, however, that risk and risk aversion must be incorporated into decision models to get useful results. Indeed, it is the contention here that for many types of decision making in agriculture, uncertainty is important but risk is not.

To assess the effect of risk and risk aversion in decision making, one needs to specify the null hypothesis. The obvious benchmark against which to measure the effects of risk aversion is the risk-neutral model. Then one can directly determine the ability of a model that incorporates risk aversion to explain actual behavior by comparing it with the risk-neutral model and using a standard goodness-of-fit test.

MODELING BEHAVIOR UNDER UNCERTAINTY

The traditional approach of estimating utility functions based on one-shot gambles in money was rejected on a priori grounds. First, the existence of a von Neumann-Morgenstern utility function in one-period money is in general inconsistent with the more widely accepted axioms of consumer choice (Spence and Zeckhauser, 1972). Second, the methods usually employed to estimate utility functions seem inappropriate for small farmers, especially where critical minimum target levels of income exist. Instead, it was assumed that if farmers are especially averse to low levels of income, their behavior can best be described by safety-first rules of thumb. The three safety-first rules surveyed in Day et al. (1971) are Roy's (1952) safety principle, Charnes and Cooper's (1959) chance-constrained programming (CCP), and Kataoka's (1963) safety-first principle. The safety principle involves minimizing the probability α that some objective function (typically profits) falls below a specified disaster level d. This is equivalent to maximizing expected utility for

$$
\begin{aligned}
U &= a \text{ for } \pi < \bar{d} \\
&= b \text{ for } \pi > \bar{d}
\end{aligned}
$$

where $a < b$.

The safety-first principle calls for

$$\begin{aligned} &\text{maximizing} \quad \bar{d} = F^{-1}(\bar{\alpha}) \\ &\text{subject to} \quad Pr(\pi < \bar{d}) < \bar{\alpha} \end{aligned}$$

where F^{-1} is the inverse of the cumulative frequency distribution. This principle may be useful where it seems more natural to identify a fixed confidence level $1-\bar{\alpha}$ than to identify the critical minimum income level. For normal distributions, safety first is equivalent to maximizing a utility function of the form $\mu - k\sigma$, where μ and σ are the mean and standard deviation of profits respectively and k is a constant. For example, if $\bar{\alpha} = 0.16$, then $k = 1$ so that indifference curves in the mean, standard-deviation plane are straight lines with slope equal to one (Roumasset, 1971).

Both the safety principle and the safety-first principle are not particularly appealing when interpreted in their utility versions. Given the preconceived notion of farmers being averse to falling below a particular income level, chance-constrained programming seems more appropriate. Under CCP the objective function π is maximized subject to a "chance constraint" of the form $Pr(\pi < \bar{d}) \leq \bar{\alpha}$. That is, risk aversion in this model takes the form of rejecting any frequency distribution with an unacceptably high chance of failure. Once this amount of security is ensured, the decision maker is assumed to be risk neutral regarding his choice among remaining distributions.

Many farmers, however, may find themselves in a situation wherein no available production technique satisfies the chance constraint. In such a case, a reasonable rule of thumb would be to come as close as possible to satisfying the chance constraint; i.e., when the chance constraint is violated for all techniques, switch to the safety principle. This composite model has been characterized using lexicographic ordering by Roumasset (1971) and is called LSF_1. The lexicographic formulation has the additional advantage of prescribing how other "ties" shall be resolved, e.g., in the case where there is no unique minimum-risk technique. Lexicographic ordering is formally a "full-optimality" model; i.e., it prescribes a unique and complete preordering. At the same time it can be used to give an unambiguous representation of Simon's (1966) notion of satisficing according to hierarchial objectives and economizing on decision costs (Encarnación, 1965).

In an alternative model LSF_2, CCP was combined with the safety-first principle. If the risk constraint is fulfilled, LSF_1 and LSF_2 predict the same choice. When the constraint is violated, LSF_2 is roughly equivalent to comparing frequency distributions according to their certainty equivalents, where the risk premium increases as α decreases.

RESULTS FROM THE PHILIPPINES

The study area chosen was comprised of rice farms in four Philippine *barrios* (see Roumasset, 1976). All farms were irrigated and almost all were using high-yielding varieties (HYVs). However, there was substantial variation in the amount of nitrogen fertilizer applied. Thus the decision studied was selection of the amount of nitrogen fertilizer applied per hectare.

To determine predictions of LSF_1, LSF_2, and the risk-neutral model, a stochastic production function was estimated, of the form

$$Y_{ij} = U_i(a_j + b_j N - c_j N^2)$$

where Y_{ij} = yield per hectare in the ijth state of the world
U_i = 1 minus the percentage of crop damage in the ith damage state
N = nitrogen per hectare

and where a_j, b_j, and c_j are parameters of the production function at the jth level of solar radiation.

Cross-section data from farmers' fields is a notoriously bad source for estimating production functions due to measurement errors (e.g., amount of nitrogen fertilizer used per hectare) and left-out variables (e.g., soil conditions). On the other hand, experimental data has limited relevance for production relationships on farm fields. It seems natural to exploit both data sources, using each for its own comparative advantage. Accordingly, experimental data was used to estimate the b_j's and c_j's, and evidence from the farm areas studied was used to estimate the a_j's and U_i's. Details of the estimation procedure are reported elsewhere (Roumasset, 1974).

The risk-neutral solution was computed for each farmer according to the first-order condition

$$\partial Y_e / \partial N = P_e$$

where $Y_e = \Sigma_j \Sigma_i Y_{ij}$
$P_e = [S_I(1 + i)P_N]/S_O P \equiv$ effective price of nitrogen
S_I = tenant's percentage payment for fertilizer
S_O = tenant's percentage share of output
i = tenant's interest rate (per cropping season) on fertilizer loans
P_N = price of nitrogen fertilizer per kilogram
P = farm gate price of rice per kilogram

The nitrogen levels that maximize expected profits per hectare for frequently observed effective prices are given in Table 4.1.

Table 4.1. Effective price ratios for representative tenancy-credit regimes and optimal fertilizer rates

Regime	S_O	S_I	i	P	P_N	$\frac{S_I(1+i)P_N}{S_O P} \equiv P_e$	N^{m*}
Biñan 1	1/2	1/2	0	0.50	1.30	2.60	70.3
Biñan 2	1/2	1/2	100%	0.50	1.30	5.20	59.0
Marayag 1	2/3	2/3	12%	0.50	1.50	3.36	60.3
Marayag 2	2/3	2/3	50%	0.50	4.50		50.9
Marayag 3	2/3	2/3	100%	0.50	1.50	6.00	38.6
Marayag 4	2/3	1	100%	0.50	1.50	8.96	14.3
Hindi 1	3/5	3/5	10%	0.44	1.54	3.85	41.4
Hindi 2	3/5	3/5	50%	0.44	1.54	5.25	30.0
Hindi 3	3/5	3/5	100%	0.44	1.54	7.00	11.1
Hindi 4	3/5	3/5	200%	0.44	1.54	10.50	0.0

*$N^m \equiv$ kilograms of nitrogen per hectare that maximize expected profit to the decision maker

Risk of disaster is defined as

$$Pr(\pi_{ij} < \bar{d})$$

where $\pi_{ij}(N) = PY_{ij}(N) - C(N)$
$C(N) =$ production cost per hectare

and where the probability distribution of Y_{ij} is based on the observed frequency distribution of U_i. For the 67 sample farmers, $\bar{d}$ was measured as the risk-sensitivity index,

$$RSI = \frac{HE + EE + UD - (OFI + LA + S + EL)}{Ha}$$

where $HE =$ expenses for household necessities in the past year
$EE =$ anticipated expense for sending dependents to elementary school
$UD =$ urgent debts (the consequences of nonpayment are greater than 100 percent interest per year)
$OFI =$ anticipated off-farm income (estimated from the past year)
$LA =$ liquid assets
$S =$ savings
$EL =$ amount of emergency loans obtainable at less than 100 percent interest per year
$Ha =$ anticipated number of hectares planted to rice in the coming year

That is, the risk-sensitivity index represents the profit per hectare needed to avoid the necessity of the farm family selling some part of their nonliquid assets.

Computer algorithms were written and used to determine predicted nitrogen levels for all 67 farmers under LSF_1 and LSF_2, as well as the risk-neutral solution. The results are shown in Table 4.2.

We are now in a position to evaluate the importance of risk in fertilizer decisions of farmers. A natural way to compare the explanatory power of the alternate choice models is to separately regress actual nitrogen per hectare used on each of the predicted values under the various models and to compare the coefficients of determination for all regression. The regression results are presented in Table 4.3.

We may reasonably assume that the constant term in these regressions represents the mean effect of left-out variables. There is no apparent reason to believe that the underlying functions have nonzero intercepts or that there are any systematic biases in errors of measurement. Specifically, we assume that the constant term represents a learning lag. As noted by Schultz (1975), the size of the learning lag presumably depends on the rate of change of the optimum levels of inputs. The fact that the lag in the risk-neutral model is moderately large (roughly ⅓ of the average optimum input) is perhaps due to the dynamic agricultural conditions during the study period. Not only were farmers using released varieties (e.g., IR20) but the prices of rice and fertilizer had also recently undergone substantial change.

Ranking the models according to R^2, the risk-neutral model performs the best, followed by LSF_2 and then LSF_1. Since the differences are not significant, these results do not prove that farmers are risk neutral or even that risk neutrality is necessarily a better description of sample farmers' behavior in general than are the *LSF* models. The results do show that supplementing the risk-neutral model with an additional concern for security does not improve the model's explanatory power. This may be partly because farmers were not particularly averse to risk; but it is probably due more to the fact that, for the techniques under consideration, risk was inversely proportional to expected profits. That is, roughly speaking, models that minimize risk tend to predict the same technique as models that maximize profits. We have not shown that Filipino farmers are risk neutral. What has been shown is that risk aversion, if it exists, is irrelevant to the demand for nitrogen fertilizer.

Fortunately for policy purposes it is not necessary to know whether farmers are risk averse or not. If risk-averse farmers choose the same techniques as risk-neutral ones, risk cannot possibly be a source of market failure.

Generality of the Results

To what extent can we generalize the result that measuring risk and risk aversion did not improve our ability to explain farmer behavior in this par-

Table 4.2. Actual and predicted nitrogen levels under alternative decision models

Farmer	P_e	RSI	N	N^m	LSF_1	LSF_2
B1	5.2	700	34.0	59.0	56.9	45.3
B2	2.6	−500	78.0	70.3	70.3	70.3
B3	2.6	−500	54.0	70.3	70.3	70.3
B4	5.2	−500	22.0	59.0	59.0	59.0
B5	2.6	−500	42.0	70.3	70.3	70.3
B6	2.6	500	56.0	70.3	71.5	62.5
B7	5.2	−500	52.0	59.0	59.0	59.0
B8	2.86	0	38.0	69.2	69.2	69.2
B9	2.6	−500	90.0	70.3	70.3	70.3
B10	5.2	−150	36.0	59.0	59.0	59.0
B11	5.2	−500	79.0	59.0	59.0	59.0
B12	2.6	−150	112.0	70.3	70.3	70.3
B13	2.6	−500	70.0	70.3	70.3	70.3
B14	2.6	−150	105.0	70.3	70.3	70.3
B15	2.86	−500	26.0	69.2	69.2	69.2
B16	2.6	−500	45.0	70.3	70.3	70.3
B17	2.6	−500	75.0	70.3	70.3	70.3
B18	2.6	−150	102.0	70.3	70.3	70.3
B19	5.2	−500	33.0	59.0	59.0	59.0
B20	2.6	800	75.0	70.3	68.5	62.5
B21	2.6	550	56.0	70.3	70.3	62.5
B22	2.6	−150	45.0	70.3	70.3	70.3
B23	2.6	−500	60.0	70.3	70.3	70.3
B24	2.6	600	56.0	70.3	70.3	62.5
B25	2.6	−500	60.0	70.3	70.3	70.3
B26	5.2	0	34.0	59.0	59.0	59.0
B27	2.6	200	45.0	70.3	66.0	62.5
M1	5.57	−999	15.0	44.1	44.1	44.1
M2	5.6	−999	0.0	43.9	43.9	43.9
M3	6.0	700	30.0	40.8	65.7	0.0
M4	6.75	300	20.0	52.5	58.7	0.0
M5	8.1	−999	12.7	26.4	26.4	26.4
M6	8.1	700	15.0	46.6	46.6	0.0
M7	3.36	− 50	23.3	61.4	18.2	0.0
M8	5.6	−100	15.3	48.5	0.0	0.0
M9	3.36	0	25.0	61.4	39.2	0.0
M10	4.5	−750	0.0	52.5	52.5	52.5
M11	4.5	700	0.0	52.5	52.5	0.0
M12	9.0	700	0.0	17.3	17.3	0.0
M13	10.0	700	0.0	9.5	9.5	0.0
M14	3.36	660	25.0	61.4	74.9	0.0
M15	3.36	400	16.7	40.8	59.3	0.0
M16	5.1	700	30.0	47.8	47.8	0.0
M17	3.36	−999	20.0	61.4	61.4	61.4
M18	9.0	570	8.7	17.3	17.3	0.0
M19	6.0	700	53.3	40.8	40.8	0.0
M20	5.4	−999	13.3	46.6	46.6	46.6
H1	10.5	49	0.0	0.0	0.0	0.0
H2	7.0	400	0.0	14.4	14.4	0.0
H3	3.85	500	0.0	43.5	43.5	0.0
H4	7.0	400	0.0	14.4	14.4	0.0
H5	7.0	−999	0.0	14.4	14.3	14.3
H6	3.85	64	0.0	43.5	16.2	0.0
H7	7.0	500	0.0	14.4	14.4	0.0

Table 4.2. *(continued)*

Farmer	P_e	RSI	N	N^m	LSF_1	LSF_2
H8	5.25	−116	0.0	30.6	30.5	30.5
H9	3.85	−999	0.0	43.5	43.5	43.5
H10	5.25	122	0.0	30.6	36.5	0.0
H11	4.55	222	0.0	38.7	51.9	0.0
H12	7.0	−999	0.0	14.4	14.3	14.3
H13	7.0	−999	0.0	14.4	14.3	14.3
H14	10.5	250	0.0	0.0	27.6	0.0
H15	11.67	− 65	0.0	0.0	0.0	0.0
H16	7.0	367	0.0	14.4	14.4	0.0
H17	7.0	420	0.0	14.4	38.3	0.0
H18	10.5	460	0.0	0.0	0.0	0.0
H19	7.0	103	0.0	14.4	21.4	0.0
H20	7.0	265	0.0	14.4	24.9	0.0

Note: $P_e = S_I(1+i)P_N \div S_O P$
RSI = risk sensitivity index = $\bar{d}$
N = actual nitrogen used (kg/ha)
N^m = nitrogen that maximizes expected profit (kg/ha)
LSF_1 = nitrogen predicted by lexicographic safety-first model 1 (kg/ha)
LSF_2 = nitrogen predicted by lexicographic safety-first model 2 (kg/ha)

ticular case? At the very least it is clearly unreasonable to conclude on a priori grounds that risk is important. Such a priori reasoning is widespread among both economists and policymakers and runs something like this. Low-income farmers are necessarily risk averse due to their proximity to subsistence levels of living. "Modern," cash-intensive techniques of production are more profitable on the average than traditional techniques, but are riskier as well. Therefore, risk aversion will induce low-income farmers to use less than the amount of inputs recommended for the modern technique.

The reasoning just cited contains two highly misleading assumptions.

Table 4.3. Estimated linear relationships of actual and predicted nitrogen inputs

Regression	Constant	Regression coefficient*	R^2
N on N^m	−17.02	0.99 (9.37)	0.575
N on LSF_1	−14.97	0.95 (8.57)	0.530
N on LSF_2	4.83	0.73 (8.85)	0.546

*t-statistics in parentheses.
Note: N = actual nitrogen used (kg/ha)
N^m = risk-neutral optimum
$LSF_{1,2}$ = optimum nitrogen under the LSF models
Sample size = 67

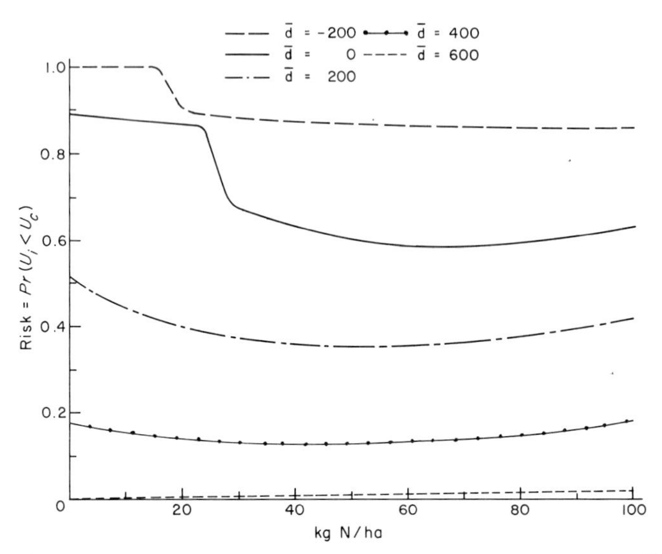

Fig. 4.1. Risk of fertilization at different disaster levels $\bar{d}$, Biñan.

First, farmers who are risk averse in the sense of having high disaster levels will not necessarily have a greater inclination to choose a low-mean, low-variance technique. Indeed farmers with high disaster levels who are following the safety principle will tend to "gamble" on the high-mean, high-variance technique (Roumasset, 1971; Masson, 1974).

The second misleading assumption is that risk necessarily increases as we move from traditional to modern techniques, thus increasing the cash intensity of inputs. Figure 4.1 shows the risk of fertilization for different levels of $\bar{d}$ and for the tenancy and interest-rate situation facing the majority of farmers in Biñan. Risk is measured as the probability that U_i is less than u_c, the "critical" percent remaining such that profits are equal to $\bar{d}$. Rather than monotonically increasing with fertilizer level, the curves are U-shaped, the minimum risk point increasing with disaster level. Figure 4.2 shows the risk of disaster where nitrogen is alternatively set at zero and at the risk-neutral optimum $N = 70$. Notice that the more cash-intensive technique is less risky than the "traditional" technique except for a very small range of disaster levels at the left side of the diagram. Moreover, the cash-intensive technique is only more risky at precisely those disaster levels where the risk constraint is not likely to be binding.

The result that adding fertilizer does not tend to increase risk can also be supported by using an entirely different set of data. Using results from nitrogen trials at experiment stations, both Day (1965) and Roumasset (1974) have shown that while fertilization increases the expected value of yields up to

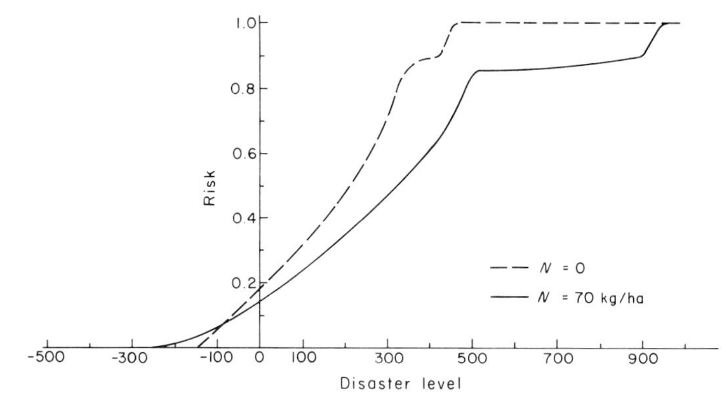

Fig. 4.2. Risk of disaster for two nitrogen levels, Biñan 1.

a point, it does not tend to increase the variance of yields. This conclusion is not applicable, however, to agricultural situations where drought is common. Experiments conducted in California and in the Philippines on nitrogen-water interaction suggest that the marginal product of nitrogen rises and then falls again as soil moisture is increased from extremely low levels to the saturation point. This implies that if soil moisture is an important source of yield variation, variance will tend to increase with nitrogen if the sample contains a large proportion of observations with extremely low water availability. If the observations are concentrated in the upper range of water availability, variance will actually tend to decrease with additions of nitrogen.

There was a tendency in several of the data sets used for negative skewness to increase with the amount of nitrogen used. It seems highly unlikely that risk would increase with the amount of nitrogen used if the expected yield was rising, variance was unchanged, and negative skewness was increasing. (The skeptical reader is invited to verify this by trying to sketch a counterexample.)

The a priori reasoning that risk inhibits the use of modern production techniques is even more misleading for the case of insecticide. Investing in insecticide is like buying insurance (albeit not a comprehensive policy); it has the effect of reducing the probability of unfavorable states of the world. Therefore, farmers who are especially averse to low income levels should tend to use more insecticide, not less.

Although existing evidence is insufficient to estimate the relationship between risk and other inputs, it seems reasonable to assume that risk does not increase with other inputs such as irrigation, weeding, and land preparation. These inputs tend to yield higher payoffs in the more unfavorable states of the world. For example, irrigation pays off most when rainfall is low; weeding pays off most when predisposing factors for a large weed population are present. Thus it appears to be more likely that these inputs will reduce risk than that they will increase risk. More formally (but less generally) we assume that if the stochastic elements are summarized in θ in the stochastic production function, $Y=f(X_1, X_2 \cdots X_n, \theta)$, then $\partial f_i/\partial\theta \leq 0$ for all i and $\partial f_i/\partial\theta < 0$ for some i, where $\partial Y/\partial\theta > 0$.

This assumption implies in turn that risk does not increase with cost to the point where expenditures on cash inputs are set at levels that maximize expected profits. Indeed this assumption is even stronger than necessary for the result. Even if some of the marginal products increase with θ (to the extent that they are complementary inputs with risk-reducing and risk-preserving inputs), risk aversion will not inhibit their use.

These conjectures cannot substitute for empirical research into the relationship between risk and expenditures in the many-input case. Their main purpose is to emphasize the unreasonableness of the a priori assumption that risk can be substantially reduced by cutting expenditures on inputs below the expected-profit maximum.

IMPORTANCE OF UNCERTAINTY

In several developing countries where high-yielding varieties of rice, wheat, and other crops have been introduced, there is a wide gap between the level of recommended inputs and the level actually used. If risk does not account for this difference, what does? One explanation is that the recommendations are not suited for the majority of small farmers. This in turn is largely because present methods of generating recommendations do not take sufficient account of uncertainty. If the risk-neutral optimum is taken as the definition of efficiency and the simple stochastic production function used above is also adopted, there still remains the substantial task of estimating the distribution of the U_i's for different localities. Experimental work is also needed to expand the simple production function, $Y_j = a_j + b_j N - c_j N^2$, to include water availability, terrain, and various characteristics of soil quality (see, e.g., Ryan and Perrin, 1974).

Positive models of farmer decisions can be improved by a better specification of the second dimension of uncertainty—learning. In the regression model introduced above, uncertainty was represented by the constant term. A more complete model would include variables that measure the extent of uncertainty about the efficient input levels due to changing varieties, prices, and environmental conditions. It would also include variables that measure a farmer's ability to make efficient decisions in the face of new information. For suggestions along these lines, see Schultz (1975).

APPARENT IMPORTANCE OF RISK

While it appears unlikely that risk increases as chemical and labor inputs per hectare increase, there has been widespread speculation that risk helps explain a farmer's apparent reluctance to adopt the set of practices recommended by government agencies. One reason is simply that recommendations have not been tailored to individual farm conditions; they have been based on data collected at experiment stations and other sites selected largely because they have level terrain, adequate irrigation, no "abnormal" soil problems, etc.; i.e., they have a lower probability of unfavorable states of the world occurring. Furthermore, experimental results under unfavorable conditions are often ignored in estimating optimum input levels. For both reasons, recommendations tend to be biased upward toward higher input levels than are efficient for the average farmer.

Since it is often difficult to criticize official recommendations and since it is no longer fashionable to assume farmers are either irrational or lazy, the hypothesis of risk aversion provides a convenient way of resolving the apparent paradox that farmers are rational but inefficient. But just as assuming that farmers are irrational is a nonexplanation of farmer behavior, so the assumption of risk aversion acts as a *deus ex machina* in reconciling fact and theory. Hiding the real explanations of behavior in a construed utility function of timeless money does little to shed light on the critical question of market failure—or on the appropriate direction of agricultural policy.

To counter the temptation to reconcile fact and theory by an appeal to risk aversion, the researcher must make a serious attempt to explain behavior with reference to more fundamental causes. For example, take the case of crop diversification. The risk-aversion explanation of diversification is that to the extent returns to different crops are not highly correlated, diversification will reduce the variance of the portfolio of enterprises and will therefore be preferred by the variance-averse farmer. We ignore here the previous point (Roumasset, 1971) that risk aversion does not imply variance aversion.

A number of other factors can explain diversification without an appeal to risk aversion. First, diversification is a device by which farmers reduce the second dimension of uncertainty by learning about new production techniques. For example, farmers commonly adopt new seed varieties on a small scale to observe their performance under local conditions. Apparent diversification may also be motivated by land productivity differences on a single farm. Because of water availability or differences in soil type, some farmers may recognize that different production techniques are optimal on different parts of their farm. Similarly, crop rotation may be mistaken for diversification.

Another motive for diversification is to make efficient use of fixed factors of production, especially capital stock and family labor. To the extent that different crops and even different varieties require different timing in various farm operations (land preparation, weeding, harvesting, etc.), diversification facilitates a higher utilization rate of fixed factors (Pope, 1975).

The last motive to be examined may be classed under the broad heading of market imperfections. Specifically, suppose the price a farmer pays to buy rice is considerably higher than the price he receives for selling it (e.g., because of marketing costs, monopoly elements, or uncertainty). Suppose also that rice is the farm family's subsistence crop and that there is an alternative cash crop with higher expected profit per hectare. If the wedge between buying and selling prices is sufficiently great, then even under certainty about yields the farm family may diversify by devoting the amount of land to rice that is needed for family consumption and using the rest for the cash crop.

Suppose now there is uncertainty about yields and the farm family is observed to allocate land so that there is a greater than 50 percent chance that the rice yield will be sufficient to meet the family consumption requirement. This appears to be evidence for risk aversion, since it is consistent with a chance-constrained programming model. In this model, the individual maximizes expected income subject to the constraint that the probability the consumption requirement (rice) is not met must not exceed some α, where income is defined as the value of output evaluated at the selling prices (Kunreuther, 1974; Kunreuther and Wright, 1974).

Note, however, that this apparent evidence of risk aversion is also consistent with the model calling for the farm family to maximize expected income remaining after taking care of its rice consumption requirement. This is accomplished by either growing it on the family farm or by buying part of it (at a higher price) with proceeds from the sale of the cash crop. This provides another example of the general thesis that many observations explained through an appeal to risk aversion can be explained just as well by a model that incorporates the relevant details of production and consumption instead of submerging them in a utility function (or the rule of thumb that acts as its substitute). The tendency that attributes deviations from the maximum expected income solution to tastes is analogous to one method for reconciling microeconomics with the demand for money. "If there was one place in theory that you would place money in order to obscure its real function as much as possible, where would you put it? In the utility function, of course" (Ross Starr, personal communication).

Another type of decision making wherein learning, not risk, is important is the incentive for farmers to adopt relatively flexible methods of production. For example, diversification in different crops and the choice of fragmented farms can be partially explained in terms of increasing a farmer's flexibility. To the extent that environmental conditions appear more favorable to one crop or plot location than another, a farmer can adjust to this situation in the middle of the growing season by allocating a greater proportion of capital and labor to that crop or location. Similarly, farmers can postpone the decision regarding the amounts and timing of various chemical inputs until they see how the crop is doing. This helps explain a common reluctance among rice farmers to apply basal applications of fertilizer.

RISK AND THE THEORY OF GOVERNMENT INTERVENTION

The important question for guiding agricultural policy is not whether risk plays a substantial role in the allocation of resources, but whether deviation from risk neutrality induces some correctable market failure. For this discussion assume that the decision maker's risk preferences can be represented by utility functions over money received in the present. It is crucial to note that the utility function of period-one money does not represent an individual's inherent tendency to gamble or to follow the safe conservative route. Rather, it is an indirect utility function derived from the individual's utility function over lifetime consumption and is dependent on the level of relative prices, especially interest rates, and the portfolio of assets held by the decision maker (Masson, 1972; Roumasset, 1976). It can be shown that as institutions for diversification and the diffusion of risk (e.g., capital and stock markets) increase, two things happen. First, the indirect utility function gets closer and closer to a straight line; i.e., both risk aversion and risk preference diminish. Second, the risk premium that represents the difference in the rate of return between "risky" and "safe" assets will also diminish (Arrow, 1964; Arrow and Lind, 1970). Both observations are two sides of the same coin, since an increasing index of risk aversion implies an increased risk premium (Diamond and Stiglitz, 1974).

The two forces that reduce risk premiums are diversification and risk sharing. As the number of assets or enterprises that an individual owns increases, the risk premium decreases. With perfect capital markets, and as long as the number of securities is not less than the number of states of the world, the competitive equilibrium is efficient even though a positive (albeit small) risk premium may exist. Thus the existence of risk aversion does not imply market failure.

The other force is risk sharing or risk diffusion. Arrow and Lind (1970) show that as ownership of a single asset is diffused over more and more individuals, the aggregate risk premium declines. The authors incorrectly extend this theorem, however, to conclude that the government risk premium must be zero (Nichols, 1974).

In Arrow's (1964) world of perfect capital markets, the benefits of a government-instituted crop insurance program would be zero. Since risk does not cause failure in that world, the institution of crop insurance would be entirely redundant to the available institutions for diffusing risk. In the second-best world, where transaction costs inhibit risk from being spread in the optimal way, crop insurance may provide some benefit, but that benefit will decrease as the risk premium becomes low and the number of substitutes for crop insurance increases.

Even with all the market imperfections that exist, a number of institutions are available to reduce risk premiums. In addition to diversification possibilities and risk-reducing inputs (e.g., insecticide), socioeconomic institutions such as sharecropping and the extended family serve to reduce the importance of risk at the margin. Ownership of semiliquid assets also reduces risk premiums, since they reduce the consequences of loss.

Furthermore, providing crop insurance is a relatively expensive undertaking, especially since an effective program must be mandatory (Mirrlees, 1974). Farmers who were already at the risk-neutral solutions may be induced to move away from them, since the mandatory insurance program changes the density functions of profits. Thus, even ignoring the costs, the allocative benefits of crop insurance may be negative. When we add the sizable administrative costs to the picture, we can confidently conclude that a crop insurance program directed at raising levels of cash inputs will have a present value considerably less than zero.

Even if there were sizable allocative gains to be had from reducing risk premiums, crop insurance is probably not the most cost-effective mechanism for doing so. Low-interest loans (say those with real interest rates less than 10 percent) reduce risk premiums close to zero (Masson, 1974) and will be relatively inexpensive to administer as long as the government can enforce a high rate of repayment. At the same time low-interest loans would offset two important areas of inefficiency. One is that farmers often face interest rates higher than the social opportunity cost of capital. The other source of inefficiency is that in dynamic agriculture tenants tend to underinvest in new, cash-intensive techniques due to lack of knowledge about their properties. Low-interest loans provide a subsidy of those inputs. The subsidy is more effective if the loans are part of a supervised credit program, which is based in turn upon a sound system of generating crop recommendations.

Finally, even where risk premiums are large and capital and other markets are imperfect, it does not necessarily follow that expected utility-maximizing decisions are socially inefficient. For example, if risk aversion is due to higher borrowing than lending rates but the differential between rates represents the competitive cost of intermediation (including enforcement), making the risk-neutral decision would be inefficient because it would entail excess use of the intermediation resources.

IMPLICATIONS FOR THE DESIGN OF TECHNOLOGY AND OTHER RESEARCH

Further research is needed to generalize the result that risk aversion does not cause substantial misallocation of resources and to demonstrate the conjecture that risk premiums are extremely small. In the meantime one should remain skeptical of results that assert the importance of risk and ask whether they are explainable on other grounds. As we have seen above, the alternative explanations include the learning dimension of uncertainty, the failure to correctly estimate risk-neutral input levels, imperfections in capital and other markets, and the motive to retain flexibility in responding to environmental conditions. It appears on both theoretical and empirical grounds that these factors are far more important to formulating agricultural policy than is risk.

The implication for designing technology is to assess the benefits of new techniques in terms of their impact on expected profits. Thus a new variety with lower susceptibility to pests and diseases is to be valued only for its ability to raise expected profits. It is not necessary to add an additional premium for that variety's ability to reduce the probability of high crop damage. In practice it seems likely that using the criterion of risk reduction will often generate the same recommendations as the criterion of expected profits. For example, varieties that have the highest impact on expected profits are likely to be exactly those that reduce the major sources of loss (measured by the mean percent damage times the probability of occurrence).

The expected profit criterion should be used at all stages of the breeding program—in deciding which characteristics to breed for, which particular varieties to retain for further research, and which varieties eventually to select for release to farmers. For uncertainty to be properly taken into account in evaluation of the potential benefits of a variety, experiments should consciously be designed to test its properties under various conditions. Thus experiments need to be conducted in different locations, at different times for the same location, and in greenhouse simulations. Then the potential benefit for a specific location can be estimated according to the weighted average of the variety's performance in the various conditions, the weights being area-specific probabilities of occurrence.

Similarly, the expected profit criterion should be used for generating recommended inputs. It may also be used to assess the expected benefits of irrigation and new farm machinery. In the latter case, it seems especially important that second-round effects on factor prices be taken into account.

The main difficulty with implementing the expected profit criterion is its complexity. Estimation of performance of a particular technique under numerous states of the world and estimation of the probabilities of the states may be sufficiently costly that subjective judgments by experts may be a superior strategy for some decisions. Thus, while advocates of complicated models of expected utility maximization or even lexicographic ordering implicitly fault the risk-neutral approach as overly simplistic, its real fault is that for many situations it is overly complex. The business of estimating probability distributions of profits for alternative techniques is often so crude, ad hoc, and based on such limited data that even estimates of the risk-neutral solution are necessarily rough. Attempting to adjust the calculation of the economic optimum for some characterization of risk preferences is a refinement that we simply cannot afford at the present time.

CONCLUSIONS

The a priori assumption that risk aversion of low-income farmers causes serious resource misallocation has no theoretical or empirical basis. The popularity of risk-based explanations of farmer behavior appears to be due to the fact that risk aversion plays the role of a *deus ex machina* in reconciling theory and reality. Explanations of behavior that simply assume it is due to some indefinable aspect of preferences neither are intellectually satisfying nor do they provide guidelines for public policy. A more productive approach is to seek the ultimate causes of behavior. The apparent importance of risk in naive models is often due to the failure to take the following into account: crop performance under the various states of the world, the learning dimension of uncertainty, the flexibility of being able to postpone decisions until some of the environmental conditions are known, efficient use of fixed stocks of labor and physical capital available to the farm family, and market imperfections. The importance of risk can only be substantiated by comparison with an appropriately specified null hypothesis that takes the above factors into account.

At the very least, proponents of government investment or foreign assistance for risk-diffusing institutions such as crop insurance should be required to demonstrate that these projects are likely to yield an acceptable rate of return. To be valid, such a demonstration would have to establish first that recommended practices have higher expected profits and risk than the practices currently employed.

Second, the demonstration would have to show that risk preferences were such that a farmer's reluctance to follow recommended practices was indeed a consequence of risk aversion. It is not sufficient to show the existence of risk aversion; one must also know its cause. For example, if buying price to the farmer is higher than selling price, he will grow more of a home-consumption good and less of a cash crop (with higher expected profits and risk) than is warranted by the expected-profit maximum. Such a farmer would appear to be risk averse; nonetheless his behavior would be efficient.

Third, it should be shown that existing institutions such as sharecropping, the extended family, and the availability of credit are insufficient to prevent risk from causing a misallocation of resources. Finally, it must be demonstrated that any efficiency gains from further diffusing risks is worth the cost of the government-sponsored institution.

Lacking such a demonstration, I would suggest avoiding expensive institutions such as crop insurance and using the risk-neutral solution as the target for public policy. Any errors in using the risk-neutral model either to generate recommendations or to explain or predict farmer behavior are likely to be minute compared to the large errors involved in estimating risk-neutral solutions and misspecifying or ignoring the learning dimension of uncertainty. The implication for designing new technology is that the efficiency benefits of potential innovations can be estimated by the impact they would have on expected profits.

COMMENT / *Mario Kaminsky*

We are moving now into the risky (or is it uncertain?) area of the importance and unimportance of risk in poor agriculture.

Let me first congratulate the author because this chapter has the two most important ingredients for a seminar of this nature: it is both bold and challenging. Roumasset has undoubtedly achieved the goals he had in mind.

The chapter structure reveals two different parts: one expands the central point that risk aversion may be unimportant for development policy considerations; the other presents empirical evidence to support such an assertion. This last undertaking is dealt with in the first part of the chapter.

In my opinion, the section attempting to clarify the question of what risk is all about is not very successful. One type of risk is said to be that returns will fall below a specific level. The probability that returns will fall below a specific level is just that—the probability that returns will fall below a specific level. It is not a risk. When talking about risk or risky conditions, such as those facing poor farmers, we are referring to events or outcomes subject to a probability distribution.

The fact that there is more than one possible outcome is not typical of decision making under uncertainty only. It is common to risky conditions and to uncertainty conditions. When there is only one possible outcome, we are happily facing a deterministic world rather than a probabilistic, stochastic one.

Uncertainty identifies itself with lack of knowledge of the relevant probability distributions. Risk identifies itself with situations where those distributions are known. These do not need to be known with precision nor to be constant under different conditions.

In a probabilistic world when a farmer makes a decision, he is in fact acting under risky conditions rather than under total uncertainty. The fact that there may be events that cannot be insured against with an insurance company is irrelevant for characterizing the situation. The farmer need only try to acquire insurance against undesirable events (e.g., returns falling below a specific level) by directing the decision-making process toward that goal. This is what a risk-averse farmer in fact does by a process that includes the approximation of the relevant probability distribution (in Roumasset's words, his "personal probabilities").

The author then attempts to assess the role of risk in agriculture. In my view what he really does is try to assess the role of risk aversion in agriculture. How this is done has already been clearly presented elsewhere (Ch. 2). The combination of *in vitro* and *in vivo* experiments constitutes the use of an ingenious device.

It is reasonable to think that risk aversion is a phenomenon increasing with the net costs of the particular practice under analysis. A few rough computations from Table 4.1 show, however, that fertilization costs amount to around 10 to 20 percent of total costs. This in turn shows that the particular sample used for estimation does not allow for dramatic conditions that may neatly lead to the typical risk-averse type of behavior.

A question, an interpretation, and a conditional warning regarding the data employed in the estimation process follow. The figures for actual kilograms of nitrogen to be used per hectare may be thought to fall from the sky, to be determined by some unknown decision-making process, or to be related to actual technical recommendations to the farmers included in the sample. If the latter proposition is more nearly true, which specific levels do those recommended levels approximate the most: levels of nitrogen (risk-neutral model optimum) or levels predicted by safety-first (risk-aversion) models 1 or 2? I am inclined to think they would be nearer to actual nitrogen levels. If this is the case, the empirical evidence is definitely biased toward the null hypothesis of the researcher, namely that risk-aversion models are poor representations of reality.

Leaving that question aside, the empirical evidence presented is relatively poor as proof of such an assertion. This judgment is based on the following:

1. First, the "betterment of fit" lies in the interval of 5 to 8 percent, and we do not know whether it is statistically significant. In any case, at a descriptive (noninferential) level it does not seem to be dramatic enough to carry much weight, especially taking into account the comment below.
2. Second, the indicator of "goodness of fit" employed is almost one of "badness of fit"; i.e., the three regression models estimated are equally poor for explaining actual behavior under farm conditions *(in vivo)*. This point reinforces the previous one in the sense that it is relatively easier to

get dramatic improvements in statistical fits, starting from a level as low as 50 percent of the total variance explained. Improving a fit by 5 to 8 percent starting in the range of 90 percent is naturally much more difficult.

3. Carlos Zulberti, whose interest and expertise in the area is known to many (guided by his ideas and data to be presented in Ch. 9), conducted an interesting experiment. He ran the same regressions for a selected subset of farmers, i.e., those supposedly more inclined to risk aversion according to the risk-sensitivity index (i.e., those showing positive scores under this indicator). The not too surprising result was that, overall, the R^2's rose from around 0.5 to around 0.8. More important, he obtained a better fit with the risk-aversion model ($R^2 = 0.79$) than with the neutral model ($R^2 = 0.76$). Whether the results of this little experiment should be taken as indicating (as the literature in the area suggests) that risk aversion plays a more important role in decisions of poor farmers than of rich farmers is something the reader may decide; I am inclined to think so, however.

Bypassing the expansions to the central point of the chapter and finally coming to the conclusions, I found them too strong vis-à-vis the empirical evidence presented.

This last point, however, must not be interpreted as stressing the shortcomings of the work, if any; it rather explicitly recognizes its attractiveness and interest, reflected in its boldness and the challenging and frank positions stated.

II

Design of Technology

JAMES H. COCK

5

Biologists and Economists in Bongoland

They sailed away for a year and a day,
To the land where the Bong-tree grows. . . .
The Owl and the Pussy-Cat, EDWARD LEAR

The biologist frequently views the economist with a certain degree of suspicion, and the economist often feels that because of this the scientist will not cooperate in giving him essential data for his work. On the basis that there is no smoke without fire, there must be some reason for this lack of understanding. In this chapter I shall not try to explain this but try to show by examples the ways in which economists can help biologists. This will be done by running through a hypothetical project showing the development of a program and the interaction between biologists and economists within it.

INITIAL SITUATION

A root crop called the Bongoyam is cultivated on about 8 million hectares in the lowland tropics and is variously estimated to feed between 100-400 million people. The crop has a six-month growth cycle, is highly drought tolerant, and is generally grown on poor acid soils. Yields are generally on the order of 10 t/ha, but record yields have been reported as high as 50 t/ha. The Bongoyam is frequently grown in association with other staples such as lentils, bananas, and soybeans. The roots are 80 percent moisture; the 20 percent dry matter contains about 85 percent soluble carbohydrates, 8 percent protein of dubious quality, and low levels of materials such as fats, fibers, etc., but very high levels of vitamins A and B_{12}. The roots are highly perishable, rotting in a few days; however, the crops are grown mostly for home consumption on the farm. The Bongoyam reaches the cities as a high-priced delicacy known in Bongoland as Bong bong (produced after a fermentation and roasting process) or in small quantities as fresh roots.

Miscellaneous reports show that the Bongoyam originated in the Pacific islands, and much folklore was associated with its use. Women of the Nóson tribe regularly ate the raw root; but this was forbidden to the men by strict taboos, and they could only eat it cooked. In most other areas of the world the root is eaten either boiled or roasted. Attempts to extract glucose and starch have been made but have frequently failed. Some animal feed trials have been done, and the Bongoyam has a surprisingly high feed value for fattening pigs. Apart from other isolated reports considered unimportant, this was the state of knowledge of the Bongoyam in 1976.

In 1977 Vespuccia, a large developed country, sent their Bing Bong team to Bongoland, the home of the Bongoyam. The team was surprised at their effusive welcome, but even more surprised at being soundly trounced in the rare sport. The President of Vespuccia was so impressed by the cordial reception given his team and the performance of the Bongolese that he decided to give a large amount of assistance to Bongoland through VAMPIRE (The Vespuccian Assistance Ministry Program for International Research and Education). Since the main product of the small country was the little-known Bongoyam crop it was decided to use the aid to support a program for its development. The decision to support this crop was somewhat political, as Vespuccia, the world's largest exporter of cereal grains, did not wish to undermine its own position as an exporter of these staples. Nevertheless, there was a real feeling of empathy by certain sectors for the Bongolese, and these philanthropic elements of the society took the only line of action they could to help their new-found friends and supported a Bongoyam research program.

PHASES

Assessment

Four biologists and four economists visited Bongoland for one year starting in late 1977. A joint survey of the situation of the Bongoyam was made. The biologists found that certain good farmers were obtaining yields of about 40 t/ha but that yields were tremendously variable due to a multitude of different unidentified diseases and pests. A tremendous number of different lines were grown, and some apparently were much better in terms of yield than others. Cultural practices were closely related to the phase of the moon; farmers planted after the harvest moon and harvested after the fisherman's moon.

The economists found that the urban population had a tremendous requirement for Bongoyams but were unwilling to buy in the local markets because much of the produce was half rotten. Bong bong was used only at a ceremony similar to the first communion and would probably have a very inelastic demand. Two large-scale factories to extract starch and glucose had failed because the price of the Bongoyam was too high and the supply was extremely erratic. A large number of small-scale starch factories working on a cooperative basis in small villages were moderately successful and had a ready market for their produce, which was used in the local papermaking industry. Local use of fresh Bongoyam as pig feed for fattening was widely practiced, but farmers said that it was bad sow food. One rather intrepid economist returned from the interior and told the strange story of a young VISIT (Voluntary International Service for Improvement of Technology) agent who had tried using the fresh Bongoyam as a pig feed. The result was that all the sows became barren and the VISIT chap puzzled and disillusioned.

The two groups returned to Vespuccia, worked, and came to the following conclusions:

1. Increasing production of the Bongoyam to a level of 25 t/ha should be possible by improved technology based on better varieties, cultural practices such as weed control, and improved planting material. (Biologists' finding.)
2. The demand for Bong bong was highly inelastic and any increase in production would lower prices and have an adverse effect on production. The demand for fresh Bongoyam of good quality was large; at a price of 50 Bong per kilo the demand would be 1 million t/yr, at 40 Bong it would be 1.5 million t/yr, but at 100 Bong it would be only 0.2 million t/yr. Because of the high production of lentils and soybeans the increase from the present 0.3 million t/yr of Bongoyam to 1 million t/yr would satisfy the main dietary deficiency of the people—a low-calorie intake. The world starch market could accept almost unlimited quantities of Bongoyam starch, as it substituted directly for cornstarch. However, to enter the world market, fresh Bongoyam would have to be produced at a maximum price of 45 Bong per kilo. Furthermore, the extraction rate would have to be increased from 50 percent to 70 percent, and transport costs to the more economical large factories would have to be reduced by improved storage methods. Due to the increase in world cassava production and the consequent decrease in price of energy sources for animal feed, the farmers would have to produce Bongoyam at 15 Bong per kilo, 30 Bong below the level for starch; hence this possibility for utilization seemed remote. Increased production of Bongoyam to the level of about 1.2 million tons would lower the price of Bongoyam to about 45 Bong/kilo (present price 72), but at that level it would stabilize if sufficient starch factories could be established. Present production costs of the Bongoyam are 500 Bong/ha, giving at 10 t/ha a gain of 320 Bong/ha. The main costs were weeding (30 percent), harvesting (25 percent), and land

preparation (20 percent). Bongoland had a disastrous balance-of-payments deficit. (Economists' finding.)

At this point the two groups came together and discussed possible action. The economists stated the need for 45 Bong/kilo of Bongoyam. A total possible production at this price was almost infinite, but the product would have to be less perishable. Below 1.2 million tons the price would increase. It was highly desirable to reach the 45 Bong level and a total production of 2 million tons, which would satisfy local needs and lead to a healthy balance of payments.

At this point the biologists said that the level of yield could possibly be raised to 25 t/ha; but better varieties, increased fertilizer, and weed control would be required, the levels of which were not known. Also, disease-resistant varieties would be needed, and it was not known if these could be found. These added factors on the present land area of 30,000 ha would raise production to 0.75 million ton. The only way to further increase production would be by using new land. The economists said this was no problem, as most farmers had excess land and rural unemployment was high. Nevertheless, at this point the whole project almost came to a grinding halt because the economists asked, "OK, what resources do you need to provide 25 t/ha technology, what inputs will farmers need, and what are your probabilities of success?" The biologists inarticulately muttered about crystal balls, putting their necks on the block, planners who were satisfied with any figure however meaningless, and went off to the local inn to commiserate. There they met Mr. Jumper, head of VAMPIRE, and he asked what the problem was. One of the biologists stated that he thought the crop had potential and they could probably meet the economists' restrictions, but they needed a team of about eight scientists and support for three or four years to see what could be done. Jumper, a farsighted man, said all right and gave them their four years; in the meantime, he gave the economists a chance to evaluate some old products that were running into problems.

Basic Biology

The four biologists, a breeder, a pathologist, an entomologist, and an agronomist went to Bongoland in late 1978. The breeder collected germ plasm and evaluated it, the pathologist and entomologist studied the diseases and pests, and the agronomist tried to collate everything and get higher yields. A biochemist was employed to improve starch extraction rates and study storage problems, a physiologist to explain the great yield fluctuations, and a soils man to look at nutrient requirements.

After four years the team reported in the following manner:

1. Yields of 20 t/ha could probably be obtained with X_1 inputs, 25 t/ha with X_2, and 30 t/ha with X_3 if varieties had certain characters that could be incorporated and certain cultural practices were used.
2. The two most serious diseases were BBB (Bongoyam bacterial blight) and superstunting. Both reduced yield potential by 50 to 75 percent. Superstunting could be controlled readily by incorporating varietal resistance. BBB could be perfectly controlled by well-developed clean seed production or by varietal resistance. However, the varietal resistance was only partial, and yields were always 25 percent below those of clean seed plots.
3. There was only one insect of major importance, the Bong hopper, but varietal resistance was excellent and could be used. However, two applications of DOT increased yields by 10 percent.
4. Storage could probably be increased from the average three days to ten days by varietal improvement or by immersing in 1 percent ethyl alcohol for 24 hours to 30 days.
5. During biochemical studies it was found that the raw root contained high levels of estrogens and progesterone, thus explaining why men could not eat raw roots, the low birth rates in the areas where women ate them, and the disillusioned VISIT chap and his barren sows. The levels of estrogen and progesterone varied tremendously among clones.
6. It was better to grow Bongoyams separately and not in association with soybeans and lentils.
7. Bongoyam produced yields of about 20 t/ha on the previously unused northeastern peninsula.
8. Yield responses to fertilizers were shown. Molybdenum requirements were universally high at 2.0 kg/ha.
9. Weed control could be achieved with four weedings at 30,000 plants/ha or with two weedings at 50,000 plants/ha. In the latter case yield was reduced by 25 percent.
10. Starch extraction could be increased to 70 percent by adding an extract of the Bongoplant leaf to the water.

Evaluation

Mr. Jumper of VAMPIRE read these findings and assigned two economists to the group. Their tasks were carefully defined as:

1. Was it feasible to produce Bongoyam at 45 Bong/kilo given the yield and input data? Which input levels were most profitable?
2. Was it worthwhile and feasible to produce clean seed for BBB? Or were varietal resistance and slightly lower yield better?
3. Shoud DOT or varietal resistance be used?

4. Was ten days of storage adequate?
5. What were the world markets for estrogen and progesterone? What were the possibilities for a Bongoyam-based industry?
6. How would monoculture of Bongoyam affect soybean and lentil production?
7. What were optimum levels of fertilizer? Did high prices of molybdenum make Bongoyam production impossible?
8. What was needed to develop infrastructure, etc., of the northeastern peninsula?
9. What was the optimum weed control-plant population combination?
10. What effect would high-protein lines have on the diet of the inhabitants?
11. Was it commercially viable to use Bongoleaf extract?

The economists studied the situation and decided that production of Bongoyam was broadly feasible but that input-output data needed refining to decide optimum levels. The biologists set to work on this. The production of clean seed was highly economical and the infrastructure for seed distribution existed. The biologists ceased to work on BBB resistance. DOT applications were very costly and did not justify the 10 percent increase in yield; biologists studied methods of incorporating more varietal resistance. Ten days of storage was adequate; work on the costly alcohol process should cease, and every effort should be made to improve natural shelf life.

The world market for estrogen and progesterone was good, but more data was needed to assess the hormone situation; two biologists and two biochemists were sent to study problems.

The trend toward monoculture of soybeans and lentils suggested little effect of changed Bongoyam technology on production. Biologists ceased work on mixed cropping. Optimum biological levels of fertilizer were rather high and economic levels very low; biologists worked on more efficient methods of application. Molybdenum cost was so low as to be negligible. Plant populations above 50,000 and even less weed control might be economic due to high costs of this operation. Biologists studied this problem to gain more data.

Higher protein levels would improve the diet of people in lower income levels. Biologists stated that no genetic variability existed for this and hence it was not possible; however, perhaps Jumper would be interested in doing another project on the high-protein Bongobean. Work on the high-protein level in Bongoyam was dropped.

The development of the northeastern peninsula was a complex problem, as no roads entered the region, no public services existed, and the population was extremely low. Nevertheless, settlements were forming along the great river Bonger; cultivation of Bongoyam could be practiced in this area in a ribbon development scheme. The starch could be transported by the well-developed river transport system to the international port of Bong Kong situated on an island in the mouth of the river Bonger. A copy of this report was sent to the government for assessment.

Development

The Bongolese government strongly supported production of Bongoyam by a large credit and technical assistance program. The new varieties yielded well and with their increased storage life were transported to the cities. As expected, prices dropped to about 45 Bong/kilo and then stabilized as large numbers of starch factories came into existence. Figures showed a Bongoyam production of 1.8 million tons. However, the situation was not as pleasant as suggested by these faceless numbers; in the poorer northern lowlands many small Bongoyam producers had gone bankrupt because their yields had not kept up with other areas. Two biologists went to the north and found that farmers were still planting at the harvest moon, contrary to recommendations, and were not using clean seed. Consequently, yields were low. Further investigation showed that since farmers were using BBB-infected seed, the harvest moon was indeed the best planting time. Furthermore, production of clean seed had not been effective in this area due to the small plots used and cross-contamination from the nearby infected plots of uncooperative farmers and from vegetable gardens. Economists found that farm size was very small; and even if yields reached the national average, farm income would be low. Biologists noted the ease with which Bongobean could be grown, and economists reported that this labor-intensive crop could give much higher per hectare income.

Rather than trying to breed BBB-resistant Bongoyam varieties that would never compete with the varieties grown in the rest of the country, it was better to grow the new Bongobean varieties in the north.

By 1984 the Bongoyam production had reached 2 million t/yr. Based on this, Bongoland was one of the world's largest starch exporters and exported birth control pills worldwide.

CONCLUSIONS

From this fable a certain philosophy of interaction between the economist and biologist becomes apparent. It is important to note the phases and functions of both groups:

Phase I: Both groups gather existing information, the biologists on possible (not probable) technical changes that can be made. Economists gather information on markets, utilization, and suggested price targets.

Phase II: Scientists gather hard data and show evidence of feasible technical achievements and alternatives.

Phase III: Economists evaluate payoff to different strategies and new possibilities. Scientists cooperate in giving more refined data when needed.

Phase IV: Continuous evaluation of technology and alternative methods of achieving goals is carried on by both economists and biologists. Effects of new technology are evaluated.

The functions of the economist are:

Phase I: Provide data on supply and demand relations. Enumerate main restrictions on marketing. Evaluate alternative uses, both for internal and export markets. Assess production costs.

Phase II: None.

Phase III: Evaluate feasibility of production with new (or to be developed) technology.
Evaluate different technologies to determine the most effective strategy.
Evaluate new market possibilities resulting from new information.
Assess effects of new technology on production of other commodities.
Determine what other changes might be advantageous.
Assess new areas for development.

Phase IV: As in phase III.
Evaluation of impact of technology.

MORAL

Economist alone makes little Bong, economist with biologist makes big Bong.

COMMENT / *Reed Hertford*

It is widely felt that the role of the social sciences, especially economics, within the international centers in Latin America (CIAT, CIMMYT, and CIP) is imperfectly defined. Economists are currently playing a variety of roles and a consensus concerning their merits has not emerged. For this reason, a paper like James Cock's—one that makes specific recommendations about what economists working in a multidisciplinary agricultural research team should do—is both welcome and timely. His proposal is far from as whimsical and light-hearted as the style in which it is written, and I hope it receives the serious attention it deserves.

Before commenting directly on his chapter, I wish to suggest some answers to what seem to be basic prior questions: Why is it that the role of economics in the international centers in Latin America has not yet been clearly defined? Why are economists still searching for their place?

First, it should be recognized that not all biological scientists are like James Cock—professionally secure, sufficiently experienced, and possessing the sort of vision that leads them to invite the participation of economists, to suggest specific roles for them in the centers' programs of biological research, and to tangle willingly with the foreign ideas of their dismal science, which more than once has threatened conventional biological wisdom.

Second, a quality dimension may have been missing on the side of economists as well: they may not have been sufficiently mature and *experimentados* to have put their necks on the chopping block with firm proposals for the role they think they should play. How else are we to explain the trend of the economics programs within the centers toward "service units," which are to respond willy-nilly—with little autonomy and independent life of their own—to the "needs" of the biological researchers? How many times have center economists been heard to say that "the biological scientists are the clientele and we respond to them?"

James Cock's position is clear in not limiting economists to this secondary role. If I interpret his suggestions correctly, he proposes that they divide their time between two sets of activities: "service research" to commodity programs and independently proposed economic research that may ultimately be more broadly of relevance to the work of the centers. In Cock's phase I, for example, it is proposed that economists gather and analyze relevant market and price information, farm-level production data, and national-level consumption and trade data independently of (albeit in parallel with) biological scientists. In this phase they would thus be acting largely as an autonomous program. In phases II and III, on the other hand, Cock suggests that economists respond directly to questions posed by biological scientists concerning the feasibility of technical goals and alternatives. I would subscribe to this division of labor, and not the one that places economists in a wholly "service research" role.

Third, there has been a problem of linkages between center economists and their counterparts in national institutions, which appears to have reinforced the problem of defining the role of economists within the international centers. The centers are resource bases for indigenous programs of agricultural research in the final analysis—support units for the national institutes and university and private sector agricultural research programs. I have observed that the forms this support should take are rapidly worked out in the case of the commodity programs of the international centers. The main reason is that the professional networks, which include center biological scientists and their counterparts in national programs, are reasonbly well developed throughout the region. There is a solid infrastructure of informal and formal communication that facilitates information flows between the international centers and the national programs and thus facilitates the definition of biological research needs. This infrastructure has been the product of a rather uniform ideology, a host of professional associations in the biological sciences, many professional journals, and the very training programs of the international centers themselves, which have further strengthened existing networks of biological scientists. Center economists claim, on the other hand, that their collegial networks are less well developed in Latin America, and I agree. Some would question that they exist at all. The upshot has been that center economists have had trouble identifying the forms their support to national programs of research should take, and this has made the job of defining their role within the centers even more difficult.

Why is the economics network less well developed? Some claim there is a shortage of human capital in economics and the social sciences in Latin America—that there are not really enough informed and well-trained people with whom center economists can link and interact. While there is some truth in this explanation, I do not believe it is the major reason. More important to my mind is that center economists in Latin America have defined their professional and ideological interests so narrowly that they have excluded themselves from the mass of Latin American social scientists and important networks that already exist in the region. Most Latin American social scientists have been reared in a theoretical and methodological tradition that is highly deductive, global ("wholistic"), and not heavily empirical. This tradition is largely alien (and of little interest) to center economists, all seven of whom in late 1975 had Ph.D. degrees from North America or Great Britain and only three of whom had a reasonably thorough understanding of major Latin American streams of thought. The center economists have a more inductive, "partial," empirical, "value-free" bent and appeal to neoclassical theories derived from northern latitudes.

In summary, the role of economists in the international centers has escaped clear definition because of some missing dimensions of staff quality and maturity, professional security, and "vision" and because collegial networks in Latin America that would be capable of assisting center economists with definitions of their programs and activities are undeveloped. An implication is that the role of economists in the international centers of this region could continue to escape clear definition, no matter how many good suggestions like James Cock's are made in the future.

Turning briefly to those suggestions, I am in general agreement with what Cock proposes but do not think he goes nearly far enough.

For example, well before Cock's phase I, it seems to me there is an extremely important role for economists, as well as other social scientists, in pointing up some of the social implications of particular research strategies. Under certain conditions, increases in production resulting from the adoption of new techniques can adversely affect human well-being. Cock gives us an example: "In the poorer northern lowlands many small Bongoyam producers had gone bankrupt because their yields had not kept up with the other areas. Two biologists went to the north and found that farmers were still planting at the harvest moon, contrary to recommendations, and were not using clean seed. Consequently yields were low." Such hardships are reinforced when opportunities for changing cropping patterns are limited by climate and soils, and institutional mechanisms of governments are too imperfectly developed to ensure that the gainers in the process of technical change can compensate the losers through income transfers, as the economist's compensation principle impractically assumes.

While being alternately aroused and incensed by this problem, the international agricultural centers have even suggested it does not exist, or that it is the responsibility of other groups. I believe, however, that the international centers must recognize, document, and anticipate these problems; feel some responsibility for them; and be in a position to consider alternative research strategies that minimize them. Both in documenting and anticipating the adverse outcomes of the adoption of new technology and in appraising alternative research strategies, I see a very important role for center economists working in Latin America. Some advances, incidentally, have already been made along these lines. There are CIMMYT's country studies of the adoption of new technologies for corn and wheat, CIAT's study of the distributional impacts of the adoption of new rice technology, and the diagnostics of potato production in Bolivia, Ecuador, and Peru which have received important encouragement from CIP. Nevertheless, much more can be done in concert with national programs of agricultural research.

In phases III and IV of his scheme Cock suggests that economists in the centers have mainly microlevel concerns and that they evaluate the central hypotheses of the biological research programs. Here too I suggest some additions. As I have argued elsewhere, the adoption of new technology (the demand more generally for new techniques of agricultural production) depends not only on microlevel profitability but also on the ways in which variables at the national and international levels are likely to impinge on acceptability and profitability through time. I refer in this regard to something seemingly as remote as U.S. agricultural policies as well as external and internal terms of trade and land tenure, credit, and input policies. Also, there are noneconomic, social, and cultural variables at the level of the individual farm which, in the modernization process, can themselves change and alter predictions based on purely economic criteria. They are all too easily forgotten—in part precisely because the models on which center economists have been so elegantly weaned are inductive and of a less global nature. Here our Latin American colleagues have much to offer, and we should call upon them.

ALBERTO VALDÉS / DAVID L. FRANKLIN

6

Evaluation of Design Parameters for Cattle Production in the Colombian *Llanos*

This study explores and illustrates a methodological approach that may be useful for the analysis of the design parameters of a technology for cattle production on extensive grazing. Particular emphasis is placed on technology developed for relatively small-scale ranches. In the context of modal farms, the methodological approach consists of utilizing computer-based simulation to identify the structural relations and to represent physical, biological, and economic risk factors that may impinge on measures of performance of proposed production technology.

Specifically, we examine the introduction of improved pastures. The performance of the new technology depends on the probabilistic nature of certain physical and biological parameters and their interaction with uncertain credit policies and price fluctuations. Stochastic dominance analysis is used to assess the preference hierarchy for the various technologies.

The setting is the Colombian eastern plains (the *llanos*), where the beef cattle enterprise is considered the forward thrust system in the process of incorporation of tropical savannahs in the allic soils of South America. The eastern plains also offer unique settlement possibilities.

These savannahs represent an extraordinary land resource estimated to be at least 300 million hectares, the majority of which are not usable for cultivation and are currently very thinly populated, both in terms of cattle and people. Land values are generally very low; labor of similar productive capacity is clearly more expensive than in crop areas elsewhere (in Colombia), and purchased inputs are expensive due to high transport costs. The scarcest physical factor is capital to buy cattle. Our area of interest is considered

The authors are grateful for the contributions of Néstor Gutiérrez, Patricia Juri, and Osvaldo Paladines, from Economics, Biometrics, and Pastures and Forages (CIAT) respectively. Valuable contributions have also been received from Rubén D. Estrada and Hernán Rivadeneira.

representative of the Colombian eastern plains, part of the Venezuelan plains, and the *campo cerrado* in Brazil.

The prevalent extensive ranching enterprises are noted for their low productivity of land and animals, the latter attributable principally to the low nutritive value (principally lack of protein) and low palatability of the native pastures during the dry season. The soils are characterized as being of good structure but of low fertility, with phosphorus and nitrogen deficiencies and aluminum toxicity. The principal climatic characteristic is the existence of a predictable dry season of 3-6 months duration (depending on the region), the deleterious effect of which is accentuated by the lack of moisture retention in the soils and the flooding of the lowlands during the rainy season.

This enormous reserve of land will contribute significantly to economic development only as agricultural research—currently under study at CIAT and some national agencies—is successful in providing technology that may raise the calving rate and liveweight gains.

Use of cut pasture and grain for intensive fattening must be discarded at the outset, due to the high cost as a consequence of the low soil fertility and high transport expense. The two approaches that remain for consideration in terms of improving productivity seem to be improved animal husbandry and/or introduction of improved pastures. However, the adoption of these new technologies, still in the design phase, may be conditioned to a large extent by the prevailing credit and price policies.

The technology is examined in the context of a credit program oriented to "small" ranches whose principal or only economic activity will be cattle. The *llanos* is a zone that includes approximately 7500 km^2, homogeneous in soils and climate; it is located in the zone of influence of the ICA-CIAT experimental station in Carimagua, Meta. Here one can now find some small ranches, each of about 250-500 ha. It has been contemplated that others may be created as a result of a credit program.

The feasibility of small ranches for this region has been analyzed in a study by Rivadeneira et al. (1976). Also appearing in that study are descriptions of the management of both cattle and pastures as well as a discussion of the technical coefficients that characterize the various management and pasture technologies. In that study it was concluded that achievement of a minimum annual cash income of approximately US$600, would require an initial complement of 36 cows and their corresponding herd. This implies that the total investment required ranges between US$9000 and US$12,000 per farm.

It is important to emphasize that the enterprise under analysis is the result of recent subdivision of more extensive ranches; consequently, certain of the management practices already adopted in other regions are still in the process of adaptation to the local conditions. In general one might say that all the traditional as well as the introduced technologies are under design. In one case the traditional technologies are designed by the ranchers themselves; in the other case they are designed by institutions like CIAT and ICA.

It is hoped that this analysis will provide criteria for design of a technology, features of which are subject to manipulation by biological research. Perhaps also it will suggest modifications to credit and price policies that might help to accelerate adoption of this technology.

SELECTION OF TECHNOLOGIES TO BE COMPARED

Research results obtained through CIAT and other institutions indicated that improvements to be realized on these infertile soils through management alone are relatively small. Thus the introduction of improved pastures based on legumes appears to be the only practical long-term strategy for achievement of significant increases in productivity on these allic soils. Table 6.1 presents the calving rate and liveweight gain increases expected as a result of introducing these technologies.

However, given the gap between the technical coefficients prevalent to date and the maximum potential for the systems without improved pastures, a question remains as to the economic potential of emphasizing adaptive research to generate local technological packages for the breeding herd that in the short run lead to calving rate increases and calf mortality decreases. One might expect to achieve this through management practices such as earlier weaning, use of minerals, improved animal health, and investment in watering places. With this emphasis on management, the "native" enterprise will continue to export the weaned steer for fattening to other more fertile regions, which are also closer to the centers of consumption.

The other strategy to be examined is the establishment of improved pastures based on legumes (specifically *Stylosanthes)* in combination with grasses. This is the key component of the technology currently under design at CIAT. A critical question here is, How can biological research lead to the design of this technology in such a manner that increases in productivity are achieved without demanding a much higher initial investment?

Table 6.1. Anticipated maximum production levels of cattle on native and improved pastures on low fertile soils of tropical America

Pasture type	Maximum calving rate	Maximum annual weight gains
	(%)	(kg/animal/yr)
Native pasture	55-60*	70-90
Sown-grass pasture	65	110-120
Sown-legume pasture	75-80	180†

Source: B. Grof et al. (1975).
*Currently estimated at 45-50 percent on the existing ranches in the *llanos*.
†Up to 1.5 AU/ha in the rainy season and 0.5 AU/ha in the dry season.

CRITERIA TO EVALUATE PERFORMANCE OF NEW TECHNOLOGY

Internal Rate of Return and Cash Flow

The internal rate of return to total investment (IRR) on the farm as a whole has been used as the principal measure of performance of the technology. In contrast to the rate of return to owned capital, the former reflects the profitability of a given technology independent of the source of funding and thus of the stimulus provided by credit subsidies.

No optimization criterion is used in this study. Were we to optimize, the objective function might consist of maximizing present value of net worth at the end of the planning horizon, given the family consumption function. In computing the rate of return, benefits include the salvage value of the improved pastures and other improvements plus the capital value of the herd at the end of year 25. For an already established enterprise, we must compare different cases according to their marginal rate of return, due to transformation of the traditional system to one more intensive in the use of capital (such as the improved pasture-production system).

With the definition of net benefits used for the estimation of rate of return, net benefits should cover not only a return to owned capital but also a return to unpaid family labor and entrepreneurial skills that would approximate their opportunity costs.

Since we are dealing with relatively small enterprises that are not well integrated into the capital markets and thus have productive opportunities but no market opportunities for their capital, the advantage of using internal rate of return rather than present value is avoidance of the rather arbitrary predetermination of a particular discount rate as the opportunity cost of money. However, for methodological purposes, we have also computed present values, using the rate of return computed for native technology as the highest possible alternative use of that capital in this zone.

On the other hand, the use of internal rate of return presents us with the methodological problem of multiple roots that can be computed when there is more than one sign change in the flow of funds across time (Hirshleifer, 1970). To eliminate the multiple-root problem, we generate only one solution—forcing the program to liquidate part of the herd in the years in which negative flows would have occurred beyond the first year of operation. Because we are attempting to predict adoption of the technology by farmers, no attempt is made to compute a social rate of return, adjusting for possible discrepancies between market and shadow prices.

A consideration we have left open in the case of the small producer (usually presumed to be risk averse) is a question as to whether an annual minimum (or subsistence) income might not be a more important criterion with which to evaluate the technology. However, the evaluation of technology with respect to a present minimum income level presents methodological difficulties beyond this illustrative effort. In particular, the establishment of a minimum income would force more explicit statements about the utility functions and risk preferences and, in general, about the consumption function of producers than we are prepared to make for small farmers, since very few as yet exist in the area. The relatively small number of general assumptions needed about the utility function of the decision maker is one of the principal motivations for using the stochastic dominance analysis.

A further difficulty with using internal rate of return as a measure of performance of the technology is that its computation (also the case with present value) is essentially an averaging process and will mask information relative to the cash flow trajectory of the systems under study. A technology that appears preferable in terms of its "acceptable" internal rate of return results on very low cash flows over the period of the enterprise might not be viable, given the imperfections of the credit market. Thus in a less formal manner we present these cash flow trajectories generated by the simulation program to demonstrate the short term behavior of the technologies under consideration.

Risk Factors in the Beef Enterprise

Rate of return has been used as a utility-free decision rule. But many sources of risk exist in the beef enterprise in the Colombian *llanos*—of a biological and technical nature as well as economic and perhaps even social nature.

Among the productive risk factors are those concerned with cattle and pasture (e.g., mortality, reproduction problems, animal diseases, and the like) as well as other physical and biological factors that may affect the output of fodder from pasture. The latter are related to pasture establishment, productivity, and duration. Also, fluctuations of weather patterns may lead to varying productivities from year to year and (more important for the region in question) the risk of destruction of an improved pasture by fire from uncontrolled burning of native pasture.

Among the transaction risk factors to be considered are price fluctuation, transport risk, and perhaps even theft. For illustrative purposes, we have chosen to focus on calving rate and animal mortality as the risk factors associated with the animal itself. For technical risk we will concern ourselves only with that associated with pasture establishment, productivity, and duration. Price fluctuations are considered as the source of transaction risk factors.

METHODOLOGICAL APPROACH

A computer-based simulation of the development of a modal ranch is used to study the potential possibility of adoption of proposed technologies in the presence of biological and economic risk. This is accomplished by providing a number of external inputs such as prices and credit and by specifying the technology through the setting of parameters internal to the simulation model. The effort then is to examine the impact of the interaction of price (and to an extent credit) with the technical description of the technology in regard to viability and financial position of the beef cattle enterprise across time.

From early research at CIAT it is predetermined that the small ranch has a maximum capacity of 36 cows and their corresponding herd; thus the system is initiated at its maximum complement of cows. Where steer fattening is accomplished with purchased steers, the breeding herd is excluded; in this case the investment is that required by the full complement of fattening steers plus the establishment of pastures. The simulated ranches are allowed to evolve to a new steady state without reinvestment of net income. However, provision is made to allow for reinvestment if net income should become sufficiently large.

Simulation

While there are a number of alternative approaches, we have considered simulation to be most appropriate, given the dynamic characteristic of the technology under evaluation. In particular, the convenience of being able to introduce random variables and the easy interpretation of the results make simulation a very useful tool in these exploratory stages.

The simulation or mathematical model developed at CIAT consists principally of three major subsystems—the herd development subsystem, the cash flow subsystem, and the routine for computation of the internal rate of return (Valdés et al., 1977). (The program executes one run in about 20 seconds of central process time on the IBM 370/145 and 100 runs in about 5 minutes.)

The deterministic herd development simulator is called HATSIM and consists of approximately 100 equations and 11 technical coefficients for representing the production system. The principal inputs are the data for the technical coefficients, describing the technology, prices, and initial conditions on the herd inventory. The principal outputs are herd development across the 25 years, cattle inventories, income, and rates of return. The herd development subsystem computes the development of the herd as a function of original inventories, calving rates, mortality, purchases, and sales by the various age and sex categories. The cash flow subsystem is used to compute the yearly expenses, sales, and net cash position. All computations are made in constant-value Colombian pesos.

The principal methodological innovations in the herd development simulation have been incorporation of random variation in a number of parameters in the simulator and utilization of the techniques of stochastic dominance for the analysis of performance of different technologies under varying assumptions of risk. (See Anderson, 1974b; Hardaker and Tanago, 1970; and Naylor 1971).

Biological Risk Factors

For the representation of biological risk factors, calving rate is specified as an average taken from some distribution without specifying how this parameter depends on a number of technical factors related to pastures, management, and the like. Variation in this parameter occurs because there are many other nonobserved variables that affect reproduction rate and mortality; thus it seems appropriate to represent reproduction and mortality rates as an outcome of sampling from a normal distribution. We do not feel that the particular distribution is of prime importance for this study. However, as further technological research results are obtained, the specification of these parameters and their distributions will become one of the objects of biological research. For example, Evenson et al. (1977), in analyzing the genotype-environment interactions, brings the economists' notions of risk and uncertainty to bear on the problem of crop improvement research strategies.

Design Parameters

Since the principal technological innovation under consideration is introduction of legume-based pastures, we felt it appropriate to concentrate the analysis of risk on factors that may affect pasture performance. Our consultations with members of pasture groups in CIAT have led us to the following specification: the principal risk is that of pasture establishment, a pasture would require two or more years to become established, and the first six months are the "riskiest." This occurs as a result of too little or too much rain, land preparation problems, low-quality seed, accidental or premature grazing, insect and disease damage, and weed competition. In the postestablishment period the principal risk factor is that associated with the burning of native pastures, which will destroy any legume and some grass species. There is also the risk of strong weed competition in the case of a poorly established pasture.

Since little research exists to permit us to assign an explicit value for the relative impact of each of these risk factors, we have combined all to produce the pattern of pasture establishment and duration given in Figure 6.1. That representation allows us to identify the following biological and physical parameters related to pasture establishment and duration: probability of pasture establishment to a given level of productivity, duration of an established pasture, and rate of decay after a pasture begins to deteriorate. In this manner we hope to achieve an adequate generalized synthesis of some pasture parameters, singly and simultaneously. Thus we can explore the

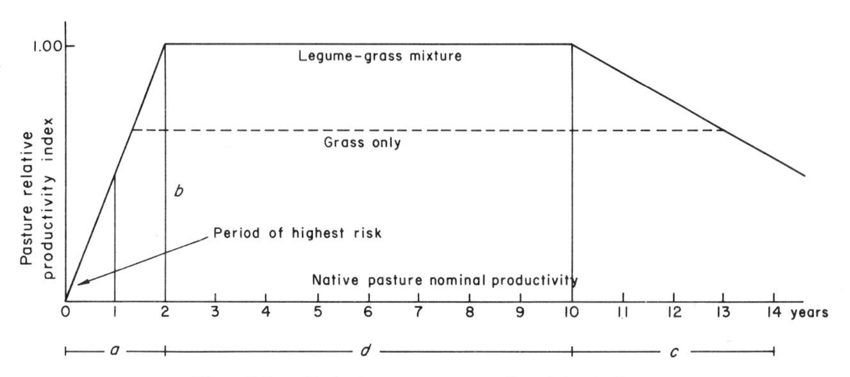

Fig. 6.1. Relative pasture productivity index.

relative merits of various conditions that agronomy, management, and breeding experts may consider as design alternatives.

Also, through biological research it may be possible to reduce the proportion of the improved pastures required within the ranch and therefore to reduce the per unit cost of improved pasture establishment. The latter may be achieved by reduction of seed prices, minimum tillage, and reduction of phosphate fertilizer requirements.

Relative Productivity of Improved Pastures

The number computed by implementing the characterization of Figure 6.1 in the simulation has been given the name of *improved relative pasture productivity index*. The index specifies the performance of an improved pasture relative to the performance of a native pasture. It serves to represent (grossly) the probabilistic elements of pasture establishment, duration, and decay.

The value determined by the trajectory outlined in Figure 6.1 could be regarded as a random variable that incorporates a number of other variables; in particular, parameter a specifies the time required for the establishment of an improved pasture to its nominal productivity, b specifies the level of productivity achieved in any one establishment, c represents the period of pasture decay, and d shows the expected duration of an established pasture given its achieved level of productivity. In the simulation we have set the value $a = 2$; the value of d is selected from a normal distribution with a mean of eight years and variance of four; c is not being utilized, since we forced the replacement of the pasture upon completion of d years; and b, which is the principal random variable in the simulation, is sampled from a normal approximation to the binominal with a probablility of 80 percent of establishment to at least 60 percent relative productivity. Since the relationship of the dynamics of pasture productivity to each of these parameters is unknown to us, only the nominal performance is specified for each technology. The relative productivity index is then used to reduce the nominal performance to its "actual" by the following formula:

$$SPP(t) = PPN + (NPIP - PPN)\, PD(t)$$

where $SPP(t)$ = a given performance parameter at time t (e.g., steer selling weight)
PPN = that same parameter for the native (or control technology)
$NPIP$ = the nominal value for the same parameter for any "improved" technology under evaluation
$PD(t)$ = the generated value of the relative productivity index

Thus the worst performance of an improved technology is that of the native technology (albeit at a higher cost).

Spectral Analysis of Long-Term Prices of Beef

In Latin America it has been the practice to manipulate price levels of beef through economic policy (Valdés, 1975). Also, it is well known that high fluctuations exist in beef prices, and it has been suggested that some beef price cycles might impinge on the "riskiness" and profitability of certain technologies. In this study we simply analyze the sensitivity of adoption of technology to variability of price and long-term price levels.

To understand how risk may come about through prices, the available Colombian time series were studied. Figure 6.2 is a graph of prices of beef on the hoof at Medellin, deflated by the wholesale price index.

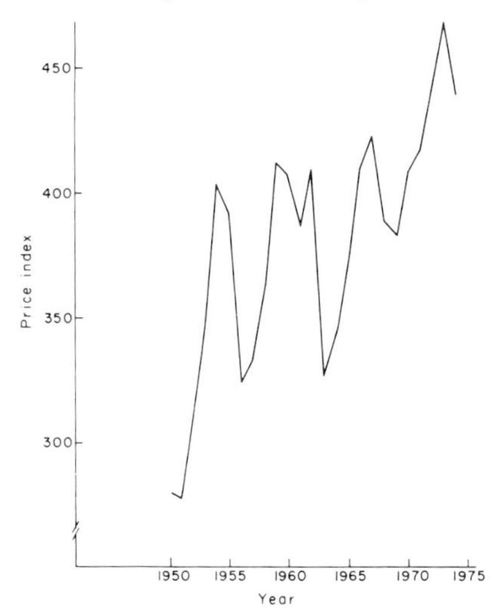

Fig. 6.2. Deflated price of cattle on the hoof in Medellin (deflated by the wholesale index).

Spectral analysis was performed in an effort to understand the variance in that time series due to periodic components (Naylor, 1971; Percival, 1974). A computer program was written to implement the methodology of Blackman and Tukey (1958). This program removes a trend line by estimating the standard regression equation $P(t) = A + BP(t-1)$, where $P(t)$ and B represent the price at (t) and trend respectively; then the autocovariance function is computed for the "trendless" time series, and finally the power spectrum is estimated from the autocovariance function. The power spectrum is used to partition the variance of the series into its periodic components. The series consisted of 24 years, which is very short; accordingly, no statistical tests were attempted on the periodic components. Maximum power was found for a period of four years; however, periods of six and eight years were also detected. These are probably not significantly different from the power occurring at a period of four years.

The following predictive model for price was thus implemented:

$$P(t) = P(O) + \beta t + \alpha\,[t(\pi + w)]$$

where β = the trend
α = the variation (one standard deviation) about this trend
w = the phase angle of the cyclic component
π = 3.1416 . . .

The values for β were set at 1.6 percent, α was set at 10 percent of $P(O)$, and w was selected randomly within the program to simulate a "start" at various points in the cycle. This means that a simulation run begins on the top of ascending or bottom of descending parts of the price cycle.

Stochastic Dominance Analysis

Stochastic dominance was considered the tool most appropriate to use for the analysis of risk, as we have no knowledge of the attitudes of individual ranchers about risk. Moreover, there are as yet few small farmers in the region. The stochastic dominance analysis makes use of frequency distributions generated by the simulation program. The rules for establishing the preference hierarchy have been stated by Anderson (1974b). That methodology was used to analyze the frequency distribution generated by 100 runs of each technological alternative. This method is based on the fact that if a sample of size n is ranked, the kth observation is an estimate of the $k/(n+1)$ fractile of the distribution from which the sample was drawn (Anderson, 1973). With these estimates of the fractiles (percentiles in our case) it is then possible to estimate the cumulative functions required by the stochastic dominance analysis.

ANALYSIS AND CONCLUSIONS

The screening of the technologies to be evaluated was accomplished in two stages. In the first stage approximately 30 cases were run under the risk-free assumption, with calving rate and mortality rate being the only random variables; the prices used were in constant real terms. The purpose of this run was to measure the sensitivity of the production system to a wide range of variation in the technical coefficients, in order to reduce the number of technologies to be evaluated. Based on these results and using the internal rate of return as the selection criterion, prototype cases were chosen for the second stage, which now incorporated the variation of price and the risk factors associated with the pasture. Table 6.2 presents the principal characteristics of the technologies selected for evaluation.

The first four cases correspond to production systems prevalent in the *llanos*. In these, calving rate, mortality, and liveweight are allowed to vary as a result of improved husbandry such as the use of minerals, vaccines, and other animal health practices. Cases 1 and 2 represent the most prevalent production systems currently in operation in the zone. It is probable that a few enterprises are represented by case 3. Based on experimental results, case 4 is known to be feasible but is probably not widely adopted in the region.

In the case of technologies based on improved pastures, there is an implied improved management to decrease the total number of hectares sown to improved pasture.

Cases 5 and 6 each have 50 hectares of improved pasture, representing approximately 20 percent of the total ranch; and management practices include minerals and animal health, leading to fattening of steers to 380 and 450 kg/animal in two and one-half years. Case 7 has approximately 8 percent and case 8 has 4 percent of the total ranch in improved pasture, leading to fattening of steers to 320 and 350 kg/animal in three and one-half years.

Results

The results for the traditional systems based on improved husbandry proved to be relatively insensitive to production risk, as shown in Table 6.3. Simple expected value analysis suggested that a moderate improvement in husbandry practices would have a better chance of adoption than more intensive practices. This is reflected in comparison of case 3 and case 4 and would suggest that, rather than research aimed at new technology, adaptive testing of practices in use elsewhere might have an appreciable payoff for the local institutions.

Stochastic dominance analysis of internal rates of return appears to be an inappropriate measure of performance for our situation. This is more of a methodological problem related to the simulations and rate of return calcula-

Table 6.2. Partial description of the technologies evaluated

System	Area		Calving	Mortality	Total initial	Annual net income††	
	Total	Improved pasture	rate*		investment†	Year 2	Year 7
	(ha)	(%)	(%)	(%)	(US$thous)		
Without improved pasture							
1. Breeding- growing	500	0	40	7-5	9.7	0.55	8.84
2. Breeding- growing	500	0	50	7-5	9.7	0.55	1.20
3. Breeding- growing	500	0	50	5-3	9.7	0.63	1.38
4. Breeding- growing	500	0	60	5-3	10.4	0.46	1.21
With improved pasture							
5. Breeding- growing	250	20	78	5-3	11.9	0.63	1.92
6. Breeding- fattening	250	20	78	5-3	11.9	0.69	1.76
7. Breeding- growing	250	8	78	5-3	10.8	0.32	1.48
8. Breeding- fattening	500	4	78	5-3	11.8	0.25	1.51
9. Fattening	50	100	...	4	11.5	2.70	2.70
10. Fattening	50	100	...	4	13.0	2.40	2.40

Note: At prices of first quarter of 1975. Initial herd of 36 cows plus younger animals and bulls, except cases 9 and 10.
*Effective rate, given a maximum value of 90 percent computed after consideration of risk factors described in the text.
†Includes value of cattle.
††Excludes value of crops produced and consumed on the farm.

Table 6.3. Internal rates of return under different systems and price assumptions

System	With price trend (1.6% p.a.)					Without price trend				
	IRR*			FIRT†		IRR		FIRT		
	Mean	St. dev.	Min.	Max.	Mean	Mean	St. dev.	Min.	Max.	Mean
					(%)					
Without improved pasture										
1. Breeding-growing	6.1	0.15	5.9	6.4	6.9	4.4	0.19	4.1	4.6	4.8
2. Breeding-growing	7.7	0.19	7.4	8.0	8.1	5.4	0.12	5.2	5.6	5.4
3. Breeding-growing	10.1	0.14	9.9	10.3	11.7	8.2	0.15	7.9	8.4	9.4
4. Breeding-growing	9.3	0.17	9.0	9.5	10.5	7.1	0.16	6.9	7.5	8.0
With improved pasture										
5. Breeding-growing	9.9	0.62	8.9	11.1	11.3	7.1	0.58	6.2	8.4	8.0
6. Breeding-fattening	10.1	0.89	8.8	11.5	11.6	7.2	0.76	5.9	8.7	8.1
7. Breeding-growing	10.2	0.35	9.6	10.7	11.4	7.5	0.59	6.3	8.2	8.4
8. Breeding-fattening	8.8	0.29	8.4	9.2	9.8	6.2	0.31	5.6	6.6	6.8
9. Fattening	...	...	...	...	...	24.0††	...	...	...	...
10. Fattening	...	...	...	...	...	18.0§	...	...	...	...

*Internal rate of return to total investment.

†Return to owned capital, with 5 percent real rate of interest charged against borrowed funds.

††Taking into consideration pasture risk.

§Taking into consideration pasture risk and double the cost of pasture establishment.

tions than a conceptual problem with either stochastic dominance analysis or use of internal rates of return as criteria.

The simulation model was forced to sell livestock prematurely to prevent repeated sign changes in the yearly cash flows. While the impact of this requirement was minor in regard to the traditional technologies, it proved to be most important in regard to technologies based on improved pastures.

The cash flow requirement was exacerbated by the need to make further disbursements for replanting a pasture that had failed to become established or that had reached its useful life. The cash reserves were generally so low that this requirement in turn forced us to set the probability of pasture establishment at levels higher than the 80 percent originally estimated. This situation indicates that the pasture establishment and duration parameters are very important for the viability of the enterprise.

The adjustments required by the rate of return calculation could be interpreted as necessary to insure viability in the "small-ranch" case, given the unreliability of credit markets. But in this fashion a methodological problem forced us to give up profitability for viability. The final consequence was that relatively little variation existed in rates of return. This result, coupled with the fact that calculation of rate of return is an averaging process, tended to mask the "riskiness" of the enterprise. Thus we are left in a quandary in regard to the trade-off between liquidity and profitability.

These problems are brought about in the real world by poorly articulated credit markets. This immediately forces us to look beyond the farm to the institutions and their role in absorbing risk through activities such as insurance programs and long-term credit. If an insurance program existed and/or capital markets for renewal of long-term credit functioned, the methodological problem we faced would not have existed. However, if the present institutional situation must be taken as given, the biological research must produce technology so good that it overwhelms the risk. In our case we must have much better than the "risk-free" performance projected by researchers.

To explore this possibility superficially, ten runs of cases 5 and 6 were made, using optimistic technical parameters for calving rate and weight gains (an average calving rate of 78 percent and weight gains of 150 kg/animal/yr). These runs raised the IRR to 9.1 percent and 10.1 percent for cases 5 and 6 respectively, indicating the need to ascertain with precision the performance parameters of the technology through extensive testing under commercial conditions.

From the technological point of view the most interesting result appears to be that of case 9—intensive fattening of steers on a relatively small ranch. This would require purchase of feeder calves from the extensive ranches and dedication of the whole ranch (50 ha) to improved pasture for fattening steers. This system differs greatly from other technologies that focus on im-

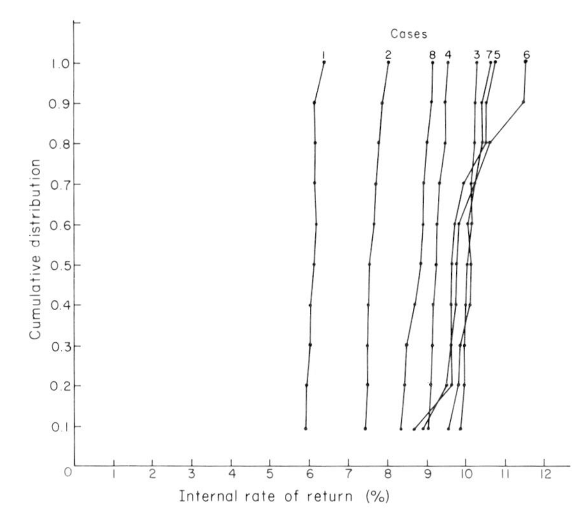

Fig. 6.3. First-degree stochastic dominance analysis.

proving the herd development parameters. These results suggest that the advantage of the legume-based pasture technology lies principally in the fattening enterprise, rather than for breeding under extensive grazing.

The stochastic dominance analysis is useful to distinguish among the available alternatives. The results for first-degree and second-degree analysis are presented in Figures 6.3 and 6.4 and can be summarized as follows:

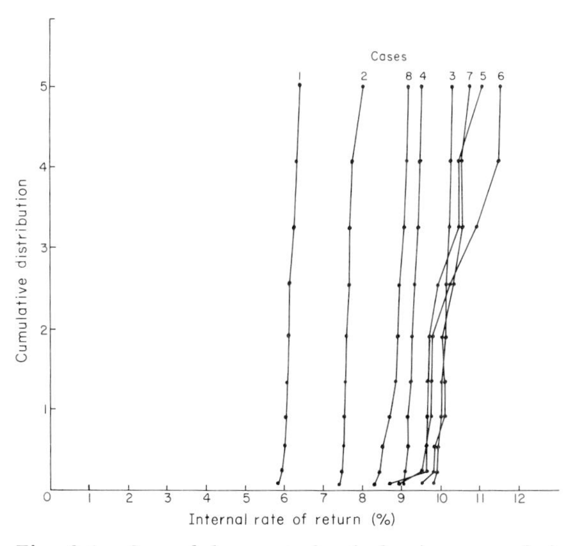

Fig. 6.4. Second-degree stochastic dominance analysis.

1. Cases 3, 5, 6, and 7 are preferred to cases 1, 2, 4, and 8; i.e., for the cow-calf operations improved husbandry is preferred to other husbandry alternatives (except fattening only).
2. Case 8 (breeding-fattening on improved pasture) is preferred to cases 1 and 2.
3. At the first-degree and second-degree analysis we are unable to distinguish between cases 3, 5, 6, and 7.
4. Stochastic dominance was not computed for cases 9 and 10. Obviously, it would have been far to the right of all the others.

Thus, as we look at the whole spectrum of evaluated technologies, it would appear that herd development parameters could now be improved through the dissemination and promotion of improved husbandry practices; legume pasture production system research should focus on developing and verifying the technical feasibility of improved pasture for fattening. Fattening appears as such an attractive alternative that even in the presence of high risk of pasture failure (Table 6.3) a much higher rate of return is computed for this enterprise. For example, our lowest estimate of case 9 (fattening only) rate of return was 18 percent, whereas our highest for all other technologies was 10 percent; the latter is capturing a price trend effect, where the former is not.

However, we see a number of limitations to the wide-scale adoption of these more intensive technologies. If the zone were to move into the fattening enterprise, we would expect the price of feeders to rise and thus reduce the margin. In this regard it should be emphasized that we took the per kilo price of feeders to be equal to the per kilo price of fattened steers; this is not common in the more fertile, traditional beef-production areas, where the relative price per kilo of fattened steers is usually higher. This peculiar condition arises in the low-fertility areas because there is ample supply of land for breeding and relatively little land for fattening. The technology under design is intended to expand the area suitable for fattening and thus should increase the demand for feeders. Hence the long-term supply elasticity for feeders becomes a critical parameter, as does that of the phosphate supply. Also, the commonly applied restrictions on credit limit the purchase of feeders for fattening, allowing the use of credit only for the breeding operation. See Scobie and Franklin (1977) for a discussion of the effect of credit restrictions on the adoption of technology.

Thus credit policies may inhibit the feasibility of this highly specialized small ranch, forcing the enterprise to be large to satisfy the requirements for inclusion of breeding stock, and to make other investments that may not be related to fattening. Furthermore, intensive use of resources may require entrepreneurial skills not covered by the net family income provision in the analysis. Finally, if this production system were to become prevalent, the price of land would also rise, and this would tend to force the rate of return to an equilibrium level.

When considering credit, another relevant element is impact of the subsidy. In this study we have used relatively low debt-to-equity ratios (between 0.25 and 0.30); thus the difference between true rates of return and the financial rate of return are low. In previous work we have used higher ratios that made the ordering of technology sensitive to subsidy. One must remain concerned that credit subsidies may lead to the adoption of technologies with low true rates of return, although high financial ones.

For our simulated price series, increase in the relative price of beef appears to dominate effects of price cycles; i.e., the expected price at the lowest level in the cycle after eight years is higher than the peak price earlier. This effect is clearly captured in rate of return calculations. In some cases the price level effects might even be as large as the technology effects (see Table 6.3). The implication of the above is that adoption of technology will be highly sensitive to price level. It has been common in South America to have consumer-oriented price policies for beef. The long-term effect of such policies may be to depress the beef sector as a whole, inhibiting the adoption of high- productivity technologies.

Another common intervention is an attempt to dampen price cycles. Our results would suggest that these efforts may have little impact on the producer with respect to technological preferences, although we would expect adjustment of inventory composition in response to changes in expected prices and cash flow requirements. In this case, terms of subsidized credit programs could possibly be made longer than the expected period, to minimize the risk of accentuating the sales effects of periodic low prices with demands for credit repayment. From a casual observation of the cash flow trajectories generated by the model (Fig. 6.5), need for an efficient market for renewal long-term credit also appears to exist.

This discussion on prices, credit, and cash flow interactions makes us think that our model should incorporate a mechanism for expected price formation and should be made to reflect the insight of recent developments in investment behavior of cattlemen in South America (Jarvis, 1974). The implied behavior rules in our model (as is usual in the normative type) are nonadaptive and thus present the risk of not predicting producer behavior, except to provide a gross ordering of technological alternatives.

The principal technology that deserves biological research attention is the development of the intensive fattening system. Our results also suggest that until the intensive system is available, better management practices may be promoted as an alternative for the improvement of the herd development parameters. For existing ranches, particularly those that would eventually incorporate finishing of cattle through the use of improved pastures, it would appear that the existing price trend tends to favor some intensification of the

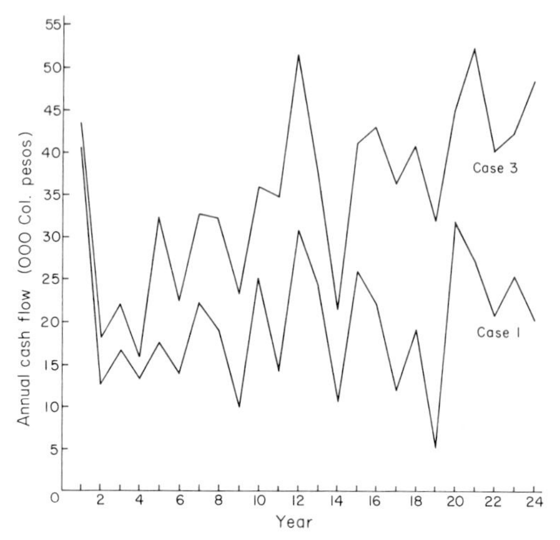

Fig. 6.5. Projected annual cash flows.

enterprise. For these ranches there still remains a question as to where to concentrate investment among management practices. This reiterates our principal conclusion with regard to improved management as a source of "risk-free" improvement of the cow-calf operation.

COMMENT / Tulio Barbosa

The chapter of Valdés and Franklin is very interesting to me. Recalling the rating by Anderson and Hardaker (Ch. 2) of various techniques to evaluate new technology, I can only hope Valdés and Franklin rate the simulation technique higher than they did! Specifically, I have the following comments.

Concerning the problem itself, I have difficulty seeing beef production as a component of a small-farm problem. Examining the data the authors present (and probably biased by my Brazilian experience) I believe a 250–500 ha ranch is not a small operation, especially if we measure size in terms of initial outlay, which would amount to US$12,000!

At least in the Brazilian case—especially considering the area known as *campo cerrado*—beef production is typically an extensive activity. The adoption of new technology in beef production there is less likely, especially in terms of pasture improvement, because it seems to conform to a situation described by Valdés and Franklin; i.e., the price of land is low and the cost of transportation is high. Even with subsidies, it seems to be more efficient for the cattleman to expand production with a relatively extensive technology.

From the point of view of small-farm operations, it is my opinion that dairying would be more consistent with small-farmer resource availability and income needs and that it should receive attention in terms of the ex ante evaluation of technology. Julio Penna (personal communication) has evaluated new technologies for dairy production in Brazil, explicitly introducing the risk element by using quadratic programming.

I conclude by their choice of simulation, that Valdés and Franklin are "risk lovers" in view of the characteristics of the technique presented by Anderson and Hardaker! The assumptions used seem reasonable to find the "stopping" point in the simulation process. However, in their conclusions, Valdés and Franklin indicate that the model should incorporate some other mechanisms (i.e., a mechanism for expected price formation) and should be made to reflect the insight of recent development in investment behavior of cattlemen in South America.

The model considers one situation for specialized farms. Under conditions of small farming, I would prefer another technique (e.g., MOTAD) in which the farm would be taken as a whole; furthermore, an optimizing criterion could have been used.

While I believe that the author's choice of credit and price policies as instrumental variables was a wise one, I am not completely convinced that the internal rate of return (IRR) is a better criterion than the present value (PV) of the net income stream to evaluate the performance of new technology. It should be emphasized that to overcome problems in estimating the IRR the authors had to resort to procedures of forcing the liquidation of part of the herd in the years in which negative flows would have occurred. I wonder whether the problems found in the stochastic dominance analysis (little variance in the estimated IRRs) could have been avoided if they had used the PV criterion instead. I appreciate that the choice of the IRR, at least in part, was dictated by the need to avoid specifying an appropriate rate of discount. However, in doing so, the chance to incorporate the time preference for different groups of farmers was lost.

JOHN H. SANDERS
ANTONIO DIAS DE HOLLANDA

7

Technology Design for Semiarid Northeast Brazil

Technological change in the agriculture of developing countries has tended to have little effect on small farmers (Bieri et al., 1972; Falcon, 1970; Gotsch, 1972). Many possible explanations for this phenomenon are amply discussed in the literature. The problem at hand is how to design new technology that will have a high probability of acceptance and utilization on small farms. For the design of this technology, research administrators and physical scientists need some guidelines and specific suggestions about its potential components.

After an initial screening process linear programming was used to evaluate a series of potential technologies for introduction on small farms in a specific region of Northeast Brazil, the Serido. We have attempted to generalize this process here.

The first step in the design process of new technology for small farmers is identification of the relevant technologies to be tested. Then decision making criteria must be specified. Using several modifications of a MOTAD programming model, we attempt to offer some insight into these steps. We "discover" a new technology that would double small-farmer incomes. We then examine the sensitivity of farm plans, income levels, and even policy recommendations. A postcript is added to summarize 1976 results from field-level testing of proposed technology resulting from the model.

POTENTIAL TECHNOLOGIES

The Tractor-Fertilizer Myth

In North America, Western Europe, and Japan agricultural development has been associated with the rapid introduction of tractors and fer-

The authors would like to acknowledge the computing assistance of Brian Hardaker of the University of New England, Armidale, N.S.W., Australia; the insightful suggestions of John L. Dillon and Pasquale Scandizzo; and the help of Antonio Clesio Thomas in our continuing struggle with the 1130 computer of the Federal University of Ceará.

tilizer. Hence there is a natural belief among many scientists and agricultural policymakers that the use of these inputs is a necessary condition for agricultural development. However, these inputs may not be available to small farms; in many cases they may not even be critical for increasing small-farmer income.

Since World War II, the use of tractors has increased very rapidly. From 1950 to 1970 Brazil increased its tractor stock from 8372 to 156,592. This increase was stimulated by various government policies subsidizing the cost. Nevertheless, in relation to the entire farm population tractors are presently used by only a very small minority, primarily the larger farmers (Sanders, 1973). In the 1980s it is unlikely that small farmers will receive the necessary credit to obtain tractors. Even if they did, it is doubtful that domestic industry could expand enough to mechanize the entire farm population over this period. Moreover, it probably is not necessary in most cases for tractors to be made available to small farmers.

In the literature on agricultural mechanization there are three principal reasons for preferring mechanical power over animal power: (1) an extremely difficult land-preparation operation, (2) "timeliness," and (3) comparative costs of animal and mechanical power. If the land preparation is so difficult that animals cannot perform at all or perform very badly, there will be a substantial yield effect from mechanical power. This occurs with certain types of soils such as those associated with the *cerrado* vegetation in the central plateau of Brazil, where small farmers shifted directly from human power to the rental of tractors. The power required for land preparation activity was greater than could be provided by animal power; hence, mechanical power eliminated a bottleneck to increase in yield and area expansion (Sanders and Bein, 1976).

The timeliness argument is that an operation must be performed rapidly to take advantage of the rain (planting) or before weather or delay can destroy the crop (harvesting). In Northeast Brazil delays in weeding also reduce yields, and adequate weeding is critical to increase in yield and area expansion (Albuquerque Lima and Sanders, 1976). If animal power and family labor are available to small farmers, this timeliness bottleneck to increase of output without mechanical power would be more relevant for larger farmers.

Finally, the low implicit family labor costs and a low opportunity cost for the land used to support draft animals would favor animal power over mechanical power on a cost basis in most of Latin America. The exception would be for large farmers, where the coordination and control problems of obtaining and operating a large labor force several times a year could raise the implicit cost of dependence upon animal and human power.

The argument for mechanization of small farmers is a critical one. In most cases a yield effect from mechanization would not be expected.

However, an agricultural engineer could identify cases in which mechanical power was necessary. Neither the timeliness nor the comparative-cost reasons would favor the use of mechanical power over animal power on small farms. However, the yield effect should be studied for particular regions. In the Serido none of these conditions existed.

Fertilizer is a very risky input for small farmers. Sources of risk are the dependence of fertilizer response upon water availability at the critical times of plant development and the large cash outlays for fertilizer purchase. Without water the response to fertilizer is limited or nil in most crops. Sorghum, millet, and other drought-resistant crops appear to be exceptions to this (see Postscript). In areas of the world without regular water availability (via irrigation or reasonable rainfall distribution during the growing season), yield insurance, or capacity of farmers to take risks, farmer interest in fertilizer would not be expected. Fertilizer also requires a large cash outlay, and small farmers in the Northeast purchase few inputs and have little access to credit markets (Patrick and de Carvalho, 1975). In determining the relevance of fertilizer for a particular nonirrigated region, the first data requirement is rainfall variability. Then it is necessary to evaluate the relationship between rainfall and yields on income and risk levels. The programming model will be used to give some of this information about the return and riskiness of fertilizer use.

In summary, tractors are not considered to be necessary for the Serido because the land-preparation operation is not very power demanding. Fertilizer use will be evaluated in the model but is expected to be a very risky activity. Therefore, in many cases biological scientists are being asked to produce a new technology for small farmers without using either of these two inputs. In the next section, we plan to make the job of biological scientists even more difficult.

Diffusion or Design?

In the 1950s the primary focus of the agricultural development strategy in developing countries was on diffusion of improved practices or new inputs through extension. This strategy was not considered to be very successful in increasing output or modernizing traditional agriculture (Hayami and Ruttan, 1971a, Moseman, 1970).

A consensus emerged that agricultural technology must be adapted to the particular conditions of the developing countries before extension would have much payoff. The international centers in the Philippines and Mexico successfully adapted new varieties of rice and wheat principally for irrigated conditions. Use of these varieties spread rapidly in the late 1960s, primarily in Asia. Their effect on small farmers can be divided into two parts.

In better agricultural regions the innovation was neutral with regard to farm size. If small farmers had land with water, they participated proportionally in the benefits of the green revolution. The introduction of the new varieties did not reach areas with poorer agricultural resources, especially those with an irregular water supply. Hence, the second effect of the introduction of the new varieties was to increase regional income disparities. To the extent that small farmers were concentrated in poorer agricultural areas in developing countries, they were made relatively worse off (Ruttan, 1976).

There are, then, two questions to consider in designing new technology for small farmers in Latin America. First, why were the new varieties diffused only to the better agricultural areas? Second, are Latin America's small farmers concentrated in the better or the worse agricultural areas? Our answers lead us to be pessimistic about the potential of traditional experiment station research for having much effect upon small farmers. The primary product of experiment station research is new varieties, whose principal characteristics are the ability to take high fertilizer levels plus having resistance or tolerance to disease and/or insect pests. The response to fertilizer is dependent upon water availability and many small farmers do not have an assured supply from either irrigation or regular rainfall. Moreover, it is assumed that farmers will be financially able to purchase fertilizer and control pests. Assumptions of access to an adequate regular water supply and credit should be questioned for small-farmer conditions in many areas.

In Latin America the landholding structure in the better agricultural areas is extremely concentrated. Small farmers tend to be located in areas with more irregular rainfall and/or inferior topography. (There are obviously many exceptions to this generalization; better empirical work on land tenure, adjusted for water availability and soil quality, is needed.) Plant breeders could be asked to breed varieties for these inferior areas; however, this strategy implies a smaller obtainable yield threshold and higher risks. (By "yield threshold" we mean the maximum possible yield obtainable with a new variety under the given environmental conditions.)

The natural tendency of research administrators and researchers is to breed for areas in which the possible yield payoff is the highest, i.e. those with better resources. The classic example of this is the choice in experimentation between irrigated, semiirrigated, and upland rice. Semiirrigation refers to river flooding in which less than complete water control is achieved. In this system taller, improved (but not floating) varieties with lower input levels to avoid lodging are preferred over the dwarf varieties with very high fertilizer use and complete water control in the irrigated system. Improvement in the semiirrigated and upland rice systems would be expected to have better income distribution consequences within agriculture than improvement in the first.

Even though researchers are concerned about the situation, in the absence of land reform, their potential contribution to the small-farmer problem is limited. One solution to this dilemma is to define the region or the product so as to force physical scientists to work for small farmers even though the possible payoffs are smaller. For example, ICRISAT has defined its regional concentration as semiarid areas; EMBRAPA's research program on sheep and goats would be expected to have much more effect on small farmers in the Northeast than their beef cattle program; CIAT's program in cassava will have much more effect on small farmers than their beef cattle program.

We have suggested to biological scientists that in many cases neither tractors nor fertilizer will be relevant and their regional concentration should be upon areas with inferior resources where small farmers are concentrated. Then how should they define the potential technologies to be studied? The researchers need to ask three questions. First, what is (are) the scarce input(s) for the specific region and clientele: land, labor, or others? For the semiarid Northeast the answer is simple. The limiting input is water, hence the most important new technologies appear to be drought-resistant crops, such as sorghum and millet, and more economical methods to conserve and use the available water. In other regions the response will be more difficult.

Second, what are the better farmers doing, and are their methods adaptable to small farmers? If, as we have argued, traditional experiment station research is unlikely to produce much benefit for inferior agricultural regions and the small-farmer clientele is concentrated there, it is necessary to take research and diffusion ideas from the best farmers in these regions. An example should make this point clear. In the Northeast a large geographical area is characterized by very similar but not identical crops, soils, and rainfall distribution. Sampling in three *municípios* indicated an extreme variation of 5 to 70 percent in the use of animals for cultivation. Economic analysis of the use of animal power indicated a reasonable internal rate of return of 35 percent (Albuquerque Lima and Sanders, 1976). The principal barrier to diffusion of this innovation over the semiarid Northeast seems to be farmer ignorance of the potential benefits and a lack of credit. Diffusion of this technology offers substantial potential to increase small-farmer income. The implication for the design of new technology is that various animal-drawn implements could be adapted for use in this region from African models or others. Agricultural policymakers should not underrate the importance of diffusion of existing technologies for agricultural areas with inferior resources. They will likely receive limited help for these regions from traditional experiment station research.

What do agricultural professionals in the Serido and similar areas recommend as potential technologies for the region? These technologies may not be adequately tested, economical to produce, or marketable; however, they give a base for further testing.

Finally, there are three desirable characteristics of new technology for small farmers. (1) The technology should have a low capital cost and require a small input from current expenses because small farmers in the Northeast have limited access to credit and they tend to avoid large cash expenditures for inputs. (2) The technology should not be unduly risky (i.e., have a large income variance), as small farmers are believed to be risk averse. (3) The new technology should not change the product mix very much unless the demand conditions for the new product(s) are also simultaneously evaluated. Not all the potential technologies that we believed appropriate were considered in our model. For further discussion see the Appendix and Dias de Hollanda and Sanders (1976).

With the above concepts in mind, we suggested the following series of principal activities:

1. Traditional crops interplanted – tree cotton, corn, and beans (cowpeas) – with traditional technology. This is the present crop combination.
2. Traditional crops with marginal changes in practices.
3. Traditional crops with more substantial changes in spacing, insect control, and other cultural practices.
4. Traditional crops with tree cotton fertilized in the first three years.
5. Traditional crops planted in strips instead of intercropped.
6. Tree cotton and sorghum planted together in strips, with one-third the area in grain sorghum.

SMALL-FARMER DECISION MAKING: INCOME MAXIMIZATION, RISK AVOIDANCE, OR SUBSISTENCE?

There is a rapidly expanding literature on the criteria by which farmers make decisions. Increasingly sophisticated mathematical and programming models have been used (Anderson et al., 1977). However, the issues here are simple to specify. How does the small farmer weight a series of objectives? How do these objectives change over time and in response to government policy measures? Before evaluating farmer behavior with different models, it is necessary to separate these three objectives conceptually.

It appears intuitively obvious that for a given risk and subsistence level any rational person would prefer more income to less. Most farmers would be expected to be adverse to higher risk levels. Obviously, among farmers (as any other group) there are gamblers who prefer higher risk levels. Individuals in this group generally either make very high incomes or go bankrupt. However, most farmers tend to avoid risk, or they need to be paid a risk premium (through higher incomes) to take increased risks. This choice of desired income and risk levels is a subjective decision influenced by many individual characteristics.

Risk is defined here as income variation. The principal sources in agriculture are climatic factors, especially rainfall (which influences yield response) and product price variation (which affects profits). In the model, historical yield and price variation are used to represent expected profitability and risk of the above types for any farm plan. Insects and diseases also contribute to yield variation, and these sources of risk would be included in the historical yield data. In practice much of this data had to be synthesized. It is preferable to give subjective probability weights to each yield and price level. However, this data is even more difficult to obtain than historic yields and prices. (See the Appendix and Dias de Hollanda and Sanders, 1976.) Farm plans that minimize risk for any income level are the most efficient. The farmer can make a subjective decision concerning his preferred plan after this most efficient set has been presented. If the researcher knew these preferences, the optimal plan could be identified.

The complication in the literature results from the subsistence objective. The argument is that small farmers will first satisfy subsistence requirements for their family and then maximize income, constrained by the highest risk level they are willing to take. This theory seems consistent with small-farmer behavior and with the failure of many programming models to predict the actual farm plan observed in the field.

A theoretical exposition of the importance of the safety-first objective in the theory of the firm is given by Day et al. (1971), while Roumasset (1974) presents an empirical application to the behavior of Philippine farmers. The sophisticated safety-first rule is that the primary objective of small farmers is to obtain a minimum income goal. Therefore, they diversify their production to obtain this goal under any weather-price conditions. They want to have a given probability level of certainty of obtaining this goal. Unfortunately, this more sophisticated safety-first rule is too vague to be refuted. As Anderson (1974b) has pointed out, "the difficulties of working with safety-first rules are the theoretical implications of discontinuous preferences at the critical levels as well as the empirical question of appropriate specification of critical levels and the probability with which they should be exceeded."

Upon closer inspection the safety-first hypothesis says little about the reasoning process of small farmers, nor does it suggest how the decision-making criteria may be modified over time or by policy measures. The subsistence objective has been misnamed. The first priority of any individual is survival. However, most small farmers do not attempt to produce all the food they consume but retain a large part of their edible production in storage for consumption until the next harvest. The subsistence objective can be considered as another type of risk avoidance. Small farmers may prefer to avoid selling all their food production at harvest time and then repurchasing food during the year. This would be logical behavior if (1) food prices fell at harvest time, (2) the risks from insect pests in storage are low (the contrary is suggested by Magalhaes Bastos and Andrade Aguiar, 1971; and Magalhaes Bastos, 1973), or (3) the risk of exploitation in the repurchase of food is high.

The latter would be the case if small farmers were constrained to purchasing necessities at a small number of stores, which could then exercise some degree of monopoly exploitation. However, this purchasing risk should decrease as roads improve, the number of stores increases, farmers obtain greater access to more stores, and more interregional trade occcurs. Hence the predominance of the subsistence requirement would be expected to be a feature of a very low-income, isolated region. Farmers might also continue producing subsistence crops because these are the best avenue for diversification against the climatic and product price risks they face.

In summary, the criteria for farmer decision making present an empirical problem with the subsistence objective as another risk constraint. By running various model specifications with and without a subsistence requirement and comparing traditional alternatives with new technologies, we hope to provide some insight into the relevant policy issues of farmer decision making and new technology choice.

OPTIMUM PRODUCTION PLANS

We have described in detail elsewhere (Dias de Hollanda and Sanders, 1976) the model, data, technologies studied, characteristics of the representative farm, and soil types. The methodology followed was the MOTAD approximation to quadratic programming (Hazell, 1971; Thomson and Hazell, 1972). In this formulation, risk is measured as the summation of the absolute income deviations from the mean rather than their variance. These income variations result from both yield and price variation over the period 1965–1973. The MOTAD model finds the minimum risk position for any specified income level. It therefore presents a series of farm plans, and the farmer specifies his preferred income-risk position. At any point within the frontier the farmer can earn the same income while taking less risk. Points above the frontier are not feasible with the given technology (see Figs. 7.1 and 7.2).

Two typical farms were defined on the basis of the SUDENE-BIRD-ANCAR data and experience in the area. (For further detail see Dias de Hollanda and Sanders, 1976.) The positions of these farms in Figures 7.1 and 7.2 are given as *M* and *N*. Their income and risk levels were obtained by forcing their observed activities and finding the income and risk levels of these production plans.

The optimum production plans for different risk-income levels without a subsistence requirement are indicated in Table 7.1. (The numbers on the effi-

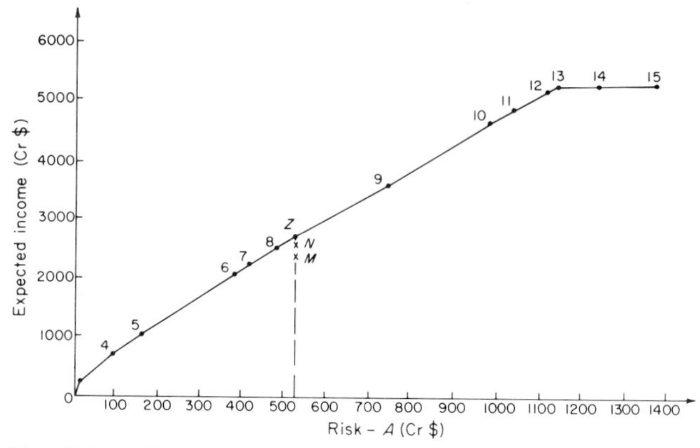

Fig. 7.1. The income-risk frontier for small farmers without the subsistence requirement and the present position of two model farms (*M* and *N*). (See Dias de Hollanda and Sanders, 1976, for an explanation of the measurement of risk variable *A*.)

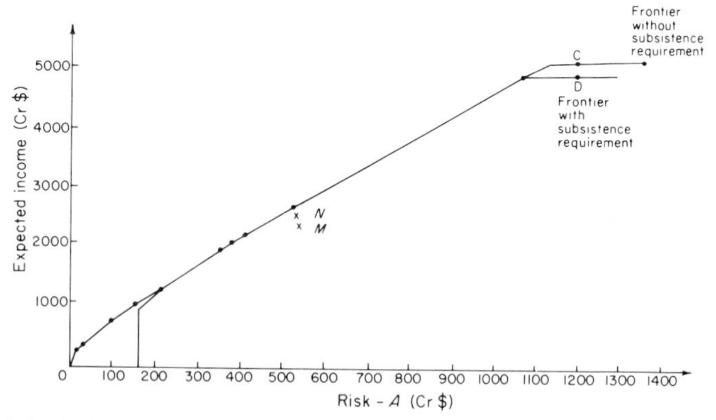

Fig. 7.2. The income-risk frontier for small farmers with and without the subsistence requirement.

ciency frontier of Fig. 7.1 refer to the optimum farm plans for different income-risk levels of Table 7.1.) Traditional technology prevails at the intermediate risk-income levels. At these levels the only difference with observed plans is on land type *A* with water, on which forage is produced rather than the two most comon crop patterns in the area—beans and sweet potatoes or beans alone (see Fig. 7.1). The distances of the representative farms *M* and *N* from the optimum plan predicted by the model *Z* are trivial, only Cr$231 and Cr$191 respectively.

Since distances from the optimum plans are so small, the model without

Table 7.1. Farm production plans for different income-risk levels without a subsistence requirement

Number	Expected income levels* (Cr$)	Area of each activity by land type (ha)					Risk levels† (Cr$)
		A	B_A	B_p	C	D	
1	285	Beans 0.16 Forage 0.14				Pasture 0.04	17
2	287	Beans 0.16 Forage 0.14				Pasture 1.34	18
3	358	Beans 0.14 Forage 0.16				Pasture 3.02	29
4	782	Forage 0.3	*Consorcio* 0.55			Pasture 7.12	103
5	1066	Forage 0.3	*Consorcio* 0.78			Pasture 22.2	159
6	2144	Forage 0.3	*Consorcio* 1.6			Pasture 22.2	382
7	2260	Forage 0.3	*Consorcio* 1.72		Pasture 2.1	Pasture 22.2	413
8	2724	Forage 0.3	*Consorcio* 2.7		Pasture 2.1	Pasture 22.2	519
9	3810	Forage 0.3	*Consorcio* 2.7	c-s 1.7	Pasture 2.1	Pasture 22.2	774
10	4806	Forage 0.3	*Consorcio* 2.7	c-s 1.7	c-s 2.1	Pasture 22.2	1011
11	4964	Forage 0.3	*Consorcio* 1.75 c-s 0.95	c-s 1.7	c-s 2.1	Pasture 22.2	1052
12	5238	Forage 0.3	c-s 2.7	c-s 0.14 *Consorcio BD* 1.56	c-s 2.1	Pasture 22.2	1130
13	5256	Forage 0.3	c-s 2.7	c-s 1.7	c-s 2.1	Pasture 22.2	1142
14	5277	Forage 0.3	c-s 2.7	c-s 1.7	c-s 1.05 *Consorcio* + fertilizer 1.05	Pasture 22.2	1239
15	5298	Forage 0.3	c-s 2.7	c-s 1.7	*Consorcio* + fertilizer 2.1	Pasture 22.2	1350

Code: Beans = cowpeas; c-s = cotton-sorghum; forage = elephant grass; pasture = native pasture; present *consorcio* = actual intercropping of tree cotton, beans, and corn; cotton-sorghum combination described in text; *consorcio* BD = same crop mix as traditionally planted in strips instead of intercropped; *consorcio* + fertilizer = traditional crop mix, with tree cotton fertilized in the first three years—tree cotton has an economic life of five years; A = area with water; B_A = old crop area; B_p = new crop area; C = worst crop area; D = better crop area.

*Mean incomes over the period 1965–1973, defined in linear programming terms as gross margins, which are gross revenues minus variable costs. Incomes specified at the basis changes.

†Mean of the sum of the absolute deviations from the mean income of the plan over the period, 1965–1973. Each combination of activities has a mean income and risk level. The computer program found the combination of activities minimizing risk for each parametrized income level.

Table 7.2. Farm production plans for different income-risk levels with a subsistence requirement

Number	Expected income levels (Cr$)	Area of each activity by land type (ha)									
		A		B_A		B_p		C		D	
1	962	Forage	0.2	*Consorcio*	1.5						
2	1097	Forage	0.3	*Consorcio*	1.5						
3	1260	Forage	0.3	*Consorcio*	1.5					Pasture	22.2
4	1985	Forage	0.3	*Consorcio*	1.5					Pasture	22.2
5	2114	Forage	0.3	*Consorcio*	1.6					Pasture	22.2
6	2260	Forage	0.3	*Consorcio*	1.7			Pasture	2.1	Pasture	22.2
7	2764	Forage	0.3	*Consorcio*	2.7			Pasture	2.1	Pasture	22.2
8	3810	Forage	0.3	*Consorcio*	2.7	c-s	1.7	Pasture	2.1	Pasture	22.2
9	4806	Forage	0.3	*Consorcio*	2.7	c-s	1.7	c-s	2.1	Pasture	22.2
10	4964	Forage	0.3	*Consorcio*	1.75	c-s	1.7	c-s	2.1	Pasture	22.2
				c-s	0.95						
11	5003	Forage	0.3	*Consorcio*	1.50	c-s	1.7	c-s	2.1	Pasture	22.2
				c-BD	0.23						
				c-s	0.97						
12	5006	Forage	0.3	*Consorcio*	1.5	c-s	1.7	c-s	2.1	Pasture	22.2
				c-s	1.2						
13	5009	Forage	0.3	*Consorcio*	1.5	c-s	1.7	c-s	1.95	Pasture	22.2
				c-s	1.2			*Consorcio* + fertilizer	0.15		
14	5048	Forage	0.3	*Consorcio*	1.5	c-s	1.7	*Consorcio* + fertilizer	2.1	Pasture	22.2
				c-s	1.2						

the subsistence requirement has successfully explained small-farmer behavior. At low-risk levels, where small farmers in the Serido were found, the new technology of the model does not offer an improvement from their present practices. At the preferred positions the representative farms *M* and *N* are earning only Cr$2335 and Cr$2469. The new technology of the cotton-sorghum *consorcio* enters the optimum plan at the income level of Cr$3810, where the associated risks are substantially higher. Farmers in the Serido area may not be aware of the possibility of producing sorghum. Field interviews indicated that most had heard of forage sorghum being produced in the area, but not grain sorghum.

Before considering the new technology recommendations at the higher income-risk levels, it is instructive to examine the results from imposing the subsistence requirement. Table 7.2 shows the optimum crop combinations forcing this requirement of a minimum 1.5 ha in the traditional *consorcio* of tree cotton, corn, and beans. Figure 7.2 combines the optimum plans for both the above cases.

Only at extremely low and high risk levels is there any difference between the two results. At the intermediate risk levels, where Serido farmers are presently, there is no difference in the optimum plans. Farmers are already essentially on the frontier at these levels. The clear implication is that the subsistence requirement is not a necessary component of the explanation of small-farmer decision-making criteria in this instance. The traditional cropping system is the most efficient plan at the low-risk income levels chosen by the farmer. At higher income levels (above Cr$4964) the optimum plan without subsistence crops does not include corn and beans. After this point there is a subsistence cost of lower income (*CD*) from continuing the production of these crops.

Returning to Table 7.1, we can note several implications for new technology. First, fertilization of cotton in the traditional interplanting system appears in the optimum plans on the worst crop area *C* but only at the highest risk levels. Second, above the income-risk level of Cr$3810 the tree cotton-grain sorghum crop combination enters the model on new crop area B_P and later replaces the traditional *consorcio* on old crop area B_A. This replacement does not begin until almost CR$5000. At an income level of Cr$5256 the substitution of cotton-sorghum is complete on three different land types. High-risk levels are necessary to earn this income, but this farm plan more than doubles the present incomes of the two representative farms *M* and *N*.

On the area with water (*A*) elephant grass was the best alternative. Various alternatives existed for land with water, but other new technologies could also be applied (see Appendix). Further experimental and modeling work may identify improved forages for the area with water (*A*) and a grass-legume combination for the better (*D*) and worst (*C*) pasture areas (B. A.

Krantz, personal communication). Grasses for areas without water would need to be drought resistant and of higher nutrient values than native grasses and the presently used forage, elephant grass.

The dominant new technology, the cotton-sorghum combination, merits closer investigation (see Appendix). Sorghum was planted on one-third the crop area in combination with tree cotton. The yields were 700 kg/ha or an equivalent of 2100 kg/ha for pure sorghum, assuming normal rainfall. Fertilization was not used and animal power was employed. These yield levels were our decision based upon discounting the experimental data available. A discount factor of approximately 30 percent was used to adjust for the difference between experimental and farm yields. Hence, experimental yields of the cotton-sorghum combination would need to be 1286 and 1000 kg/ha of sorghum to obtain 900 and 700 kg/ha under farm-level conditions. Experimental yields of pure sorghum would need to be 3858 and 3000 kg/ha to attain farm yields of 2700 and 2100 kg/ha.

The experimental yields were discounted because they generally included fertilization. The other discount factor was the standard adjustment for the better soils and management usually found in experimental conditions (Dillon, 1968). Since the yield decision was arbitrary, sensitivity analysis was done on the optimum farm plan using sorghum yields of 600 and 900 kg/ha.

At the lower yield level of 600 kg/ha the cotton-sorghum combination almost drops out of all the optimum plans. It is still found at intermediate risk levels on the inferior cropland *C*. At 900 kg/ha the cotton-sorghum combination substitutes at lower income-risk levels for the traditional crops than it does in the case of 700 kg/ha. This new crop combination of cotton and sorghum first appears at an income level of Cr$2481, which is approximately the income of the two representative farms *M* and *N*. The representative farms could shift to the new technology without taking higher risks at this yield level for sorghum. To make this product shift, farmers must know the necessary production and marketing information and be able to buy their corn, beans, and other food commodities rather than producing them.

In summary, we have estimated the profitability and risk of a new technology involving low capital costs, low current expenses, and intermediate risk levels. The new crop combination is better suited to the resource constraint of insufficient and irregular water supply than the present crop combination including corn and beans. If farmers can obtain 700 kg/ha of sorghum (2100 kg/ha of pure sorghum) in combination with tree cotton and small-farmer risk aversion can be reduced by government policy, this new technology apparently could double small-farmer income. If farmers can attain 900 kg/ha of sorghum in *consorcio* with cotton, the risk aversion of small farmers will not be as difficult a problem because this new technology enters at a lower income-risk position. In this higher yield case the primary problems would be extension of the technical production knowledge and marketing. If only 700 kg/ha can be attained, farmers will need some government help to reduce their risk aversion. Several policy instruments such as yield insurance, high levels of minimum prices, and liberal credit arrangements could be used.

According to preliminary data approximately one-half the corn produced on small Serido farms is presently retained as animal feed. In a study on the potential demand for grain sorghum in the Northeast the importance of advance contracts between farmer cooperatives and chicken ration factories has been stressed. The marketing risk of the new crop, sorghum—much of which (unlike corn) has to be sold—is believed to be the main barrier to its rapid introduction. There is no problem of inadequate demand according to Campos Mesquita et al., (1976). The problem is coordination of a marketing agreement between cooperatives and feed manufacturers or mixers. Experimental work has already identified sorghum varieties that have substantially outyielded corn under the highly variable rainfall conditions of the semiarid Northeast (see Fariz and Ferraz, 1974).

These model results are based on our best estimates from the available experimental data on sorghum in the Northeast (see Appendix and Postscript). It could be argued that the results were obvious without the programming. Clearly, sorghum should be cultivated in semiarid areas and corn and beans should be located in regions with an adequate, regular water supply. However, the next best method (after programming) to define a new technology for a specific region is to bring together the better farmers, extension agents, and researchers in the area and have them pool their collective wisdom. This method was tried in the Serido region and did not produce the same results. (Activity level 3 described earlier was chosen by this group but was rejected by the model used in this chapter; see EMBRAPA, 1974.) In retrospect this is not surprising, since without programming it is difficult to simultaneously consider risk, income, many technologies, different land types, and price and yield variation over nine years.

CONCLUSIONS

Our results indicate that present farmer behavior in the Serido can be explained as a crop diversification to protect against the risks resulting from climatic and product price variability. This is an alternative explanation of farmer behavior rather than refutation of a subsistence-first strategy. Conceptually and based upon field interviews in the region, we believe it is not necessary to include a subsistence requirement in future modeling of the region. However, development programs to encourage crop shifts of small farmers out of corn and beans must be concerned with the out-of-season availability and prices of these essential dietary components.

Results also indicate that it is possible to double small-farmer income by substituting the tree cotton-grain sorghum combination for the traditional intercropped combination of tree cotton, corn, and beans. This is a logical substitution because both tree cotton and sorghum are drought resistant. Corn and beans are notoriously sensitive to the extreme rainfall variation that characterizes the region. Future modeling and experimental work would probably lead to the identification of improved forages for semiarid areas. Thus the optimum, long-run, principal activities projected for the region appear to be tree cotton, grain sorghum, and livestock.

Special efforts would be necessary to ensure that small farmers actively participated in these projected product and input shifts. These efforts would include the provision of improved sorghum varieties rather than hybrids and restriction of subsidies for purchase of mechanical harvesters. Small farmers might then be expected to have a comparative advantage in the production of sorghum stemming from their ability to harvest without mechanization or to control the bird problem with use of family labor. Measures to reduce the increased risk of these new activities and to improve the marketing channels would also facilitate their introduction.

One major contribution of economists in the process of new technology design is to indicate data gaps in experimental and farm level work. This type of programming helps to identify promising research directions. When much of the time series data is synthetic, further experimental and farm-level testing will often be necessary. Other policy measures such as minimum prices and yield insurance can be usefully analyzed with this model for their effects on expected income, risk, and introduction of new technology.

POSTSCRIPT

After the conference in November 1975, the recommendations of this chapter were field tested in 1976 in a series of farm-level experiments in a rural development project of the Brazilian state of Rio Grande do Norte in the semiarid Northeast. The climatic conditions were adverse in that year, with almost no rain for 30-40 days during the growing season. This is known as a minidrought, and its probability has been estimated as 30 percent as compared with a full-scale drought, which has a probability of approximately 12 percent.

The results of these farm-level experiments are interesting for the corn-sorghum comparisons, the accuracy of the yield data, and the effect of fertilizer on sorghum under adverse rainfall conditions (Barbosa et al., 1976). In this type of adverse year the corn crop failed completely. The stalks, with almost no grain formation, were used as animal feed. The crop combination of cotton-sorghum, with one-third of the area in sorghum as in the model, had an average yield of 472 kg/ha for five sites in the Serido. (In a pure crop stand this would be 1416 kg/ha.) There were three other sites in which the yields were 0, 81, and 103 kg/ha. However, in these areas the soil was inferior and the experiments were planted later than they should have been.

With better rainfall conditions the sorghum would be expected to produce 1000 kg/ha, thereby raising the expected yields to 722 kg/ha. This expected yield was calculated as 0.3(472) + 0.12(0) + 0.58(1000). (Again in a pure stand this would be 2166 kg/ha.) We disagree with Ryan's comment that the sorghum yields are overstated. The continuation of these field trials in 1977 and 1978 should help resolve the data problem of the proposed crop combination. However, the past data and the 1976 results are consistent with the model data series.

Another interesting result of the 1976 field tests was that, even under adverse climatic conditions, low levels of fertilizer use were profitable on sorghum. With better rainfall conditions there should be an even greater return to fertilizer. Since other studies (Jackson Albuquerque and Sanders, 1975; Almeida Carvalho et al., 1976) have shown little fertilizer response on the traditional crops of the Northeast, it appears necessary to obtain drought-tolerant crops or varieties and then consider fertilization. Further field work is continuing on low levels of fertilization on sorghum in the tree cotton-sorghum crop combination.

APPENDIX

In this Appendix we suggest seven new technologies for the semiarid Northeast. The first two have been investigated in some detail in the research of the Department of Agricultural Economics of the Federal University of Ceará, Brazil. The other five represent research priorities and hypotheses about potentially profitable new technologies. These result from the same type of screening criteria discussed earlier—low capital cost, low current expense, low risk or little income variation between seasons, and little product shift.

In the short run the current basic *consorcio* of cotton-corn-beans would be maintained, with the introduction of sorghum on the corn area used for animal feed. In the long run corn would be phased out and the sorghum sold to ration factories or mixers. The beans would either disappear from the plan or, with increased productivity and the present very high prices, move into the areas with water (*A*). The rest of the modifications would be to increase the productivity of these activities plus some others that are complementary. These other activities include the livestock operation and small areas with water that are next to either a dam or a river. For more descriptive detail on the farming systems of the Northeast see Johnson (1971) and EMBRAPA et al. (1974).

Use of Animal Power

Presently, approximately 68 percent of the Serido farmers employ animal power; however, the rates are much lower in many other areas of the Northeast; these rates were 5 percent in one sample in Pernambuco and 28 percent in Canindé, Ceará (Albuquerque Lima and Sanders, 1976; Barbosa, 1975). Animal power is principally used for cultivation, which appears to be the principal bottleneck to increased production (Albuquerque Lima and Sanders, 1976). It is also used for transportation and for some ploughing.

There is a substantial potential for quick payoff in other areas of the Northeast through diffusion of this innovation. Under normal weather conditions the rate of return to the introduction of this input, combined with stump removal and the purchase of an animal, is 35 percent (Albuquerque Lima and Sanders, 1976). In the Serido improved implements imported and adapted from African countries could be tested under farm-level conditions.

Introduction of Sorghum or Millet

Introduction of sorghum or millet is the really high payoff innovation for the Serido. A drought-resistant, animal-feed source is long overdue. Locally selected, high-yielding varieties are now available and have been tested under farm-level conditions (see Postscript; Dias de Hollanda and Sanders, 1976; Fariz and Ferraz, 1974; and later IPA publications). These varieties substantially outyield corn under variable rainfall conditions. If advance contracts were made between chicken feed companies and farmer cooperatives, the marketing risk for grain sorghum would be substantially reduced (Campos Mesquita et al., 1976). In the Serido there already has been some production of sorghum for forage.

Even small farmers can substitute sorghum for some corn area without modifying their consumption patterns, as much of the corn in the region is used for animal feed. With some strategic government help, sorghum culture could be diffused all over the semiarid Northeast.

Two production problems still exist with sorghum—birds and insects. Control of birds plagues sorghum growers everywhere. The small farmer with his low opportunity cost for family labor may have a comparative advantage in controlling birds and harvesting sorghum. Millet would be advantageous for the substantially drier areas of the Northeast, possibly including the Serido. However, experimental research in millet is not as advanced as that in sorghum.

Improved Storage of Beans

The bean weevil *(gorgulho* in Brazil and *zabrotes* in technical terms) can completely destroy beans in storage. It can be controlled by storing beans in metal or plastic containers after treatment. The economics of control needs to be explored further.

Practices in the Small Areas with Water

Many farms in semiarid areas have access to a dam or river; this water, together with chemical or organic fertilizer and insect control, can be used to produce high-value crops such as sweet potatoes, elephant grass, rice, beans, melon, watermelon, and squash—realizing substantial productivity gains. Basic experimental and marketing research could be obtained from irrigation projects and modified for the particular characteristics of this area.

Improved Productivity of Livestock

For small farmers the principal type of capital formation appears to be the accumulation of an increased number of cattle. These farmers in the Serido produce cattle in a labor-intensive manner by putting much of the area with water in elephant grass, which is manually harvested. It may be possible to increase the yields of this grass, introduce a better grass, or introduce prophylactic measures to improve animal health.

Variety Improvement

Higher yielding varieties of tree cotton are available at the EMBRAPA Cotton Center in Paraiba. Most of these are precocious, and so involve higher risks. Nevertheless, some farmers are undoubtedly prepared to accept this, and crop insurance programs could be devised to reduce the risk through public sector support. Similarly, the improved bean varieties from the International Institute for Tropical Agriculture should be tested under farm-level conditions in the Serido and other regions of the Northeast.

Improved Use of Organic Material

Manure is commonly used in the areas with water, generally with sweet potatoes. The total production is not used, and it is not stored well. A compost heap and shading from the sun would increase its productivity.

Conclusion

The above package or some components could be expected to increase the incomes of small farmers, and favorably affect the distribution of income.

COMMENT / *James G. Ryan*

This chapter is a good example of ex ante analysis in the design of technology. The technique is not restricted to the design of technology for small farmers alone but is appropriate for all types of farms.

The major conclusions seem as follows:

1. Actual farm plans in two situations sampled by the authors very closely resembled the profits-deviation frontier (lower end). Hence in this situa-

tion new technology did not offer an improvement for obtaining additional income at no expense in terms of increased riskiness of income.

2. The addition of a subsistence requirement to the model altered the profits-deviation frontier only at the extremity; hence this is not a necessary component of the explanation of small-farmer decision making.
3. Sanders and Dias de Hollanda used the model to define in precise terms what is required of researchers to increase income of small farmers; e.g., produce a sorghum capable of yielding 2100 kg/ha of grain. This could increase incomes of small farmers by 100 percent.

Sanders and Dias de Hollanda evaluate prospective new technologies for small farmers by attempting to minimize risk, reduce capital requirements, reduce current expenses, and identify plans to generate income increases. This constructive approach is in marked contrast to some of the discussant's experiences with socioanthropologists who often castigate the green revolution for ignoring small farmers.

I question the extent to which the additional risk from altering farm plans along the profits-deviation frontier was a consequence of price or yield variability. It would be illuminating to separate these sources of variation in profit. The price and yield risk of sorghum is unknown and no doubt critically influences the Sanders-Dias de Hollanda results. Could the authors obtain a more precise estimate of the likely variability in sorghum production from data in similar agroclimatic regions of Africa and India?

The yield assumption of 2100 kg/ha of pure grain sorghum used in the analysis seems very much higher than would be expected in India where average yields are only 600–700 kg/ha. The average sorghum yield in the world, including the United States, is only about 1200 kg/ha. Since the sensitivity of the results to sorghum yield changes is rather substantial, this is a critical element in the research and could in fact lead to what may be considered a tautological conclusion, i.e., that the new technology will be successful if it is successful!

I have not seen a description of the model, data, and technologies used in the analysis. However, the benefits of parametrizing the model to examine the profits-deviation frontier for different sizes of farms, as well as other resource endowments, would be informative in answering the question of whether these prospective technologies have vastly different relevance to large and small farmers.

When considering technology for small farmers, we should not concentrate only on this group lest we swing the technological pendulum too far. There may be many technologies equally applicable to all farm sizes.

Sanders and Dias de Hollanda raise a very relevant issue concerning the extent to which the subsistence requirement on small farms is a result of the comparative advantage of growing food crops or of the farmer's perceived need for a guaranteed food supply (and a reluctance to rely on purchased food).

Discussion of the effects of shadow prices on constraints in the models examined would be useful. Solutions were generated, and it would be interesting to know whether the shadow prices have relevance in the context of the induced innovations hypothesis and its implications for research strategies. It would be especially useful to make a comparison of the shadow prices of the actual farm plans with those of new-technology plans having the same income-risk values.

The discussion of the tractor-fertilizer myth seems a trifle unrelated to the main thrust of the chapter. In semiarid tropical India, on the black soil we have found that bullocks can return to the field for sowing, cultivation, and intercultivation practices following substantial rainfall much earlier than tractors. For small farmers this timeliness is particularly advantageous.

Mittal et al. (personal communication) have shown the substantial economic benefit of improved animal-drawn implements from hundreds of experiments in farm fields. In spite of such convincing evidence, it remains a fact that the level of adoption of this technology has been disappointing. There is a definite need for further research as to whether the capital constraint is the overriding factor explaining nonadoption in this instance.

As Sanders and Dias de Hollanda point out, use of fertilizers in Northeast Brazil may be a risky enterprise (although they present little evidence of this); but if we could harvest water from excess runoff in an efficient manner and use it for supplementary irrigation to increase and stabilize yield (and perhaps raise cropping intensity), the economics of fertilizer use could be altered. This is especially so if genotypic advances occur at the same time.

If we specify a low-input, mediocre-yield technology strategy for small farmers, we could be accused by the social anthropologists of constraining all small farmers to an improved but mediocre standard of living and of failing to offer them possibilities of more dramatic income gains. This could be the ultimate irony of such a strategy, which of course would be intended to ensure just the opposite.

In less developed countries all small farmers are not necessarily concentrated in the poorer agroclimatic areas. In Asia small farmers appear to be spread across the various zones, and it would be useful if research was directed toward characterizing the distribution of small farms in the less developed countries so that this issue could be examined more critically.

PASQUALE L. SCANDIZZO

8

Implications of Sharecropping for Technology Design in Northeast Brazil

> The first and the most elementary effect of poverty is to enforce the very attitudes and behavior that make it self-perpetuating.
>
> JOHN KENNETH GALBRAITH

The purpose of this chapter is to analyze some effects of the economic dependence between sharecroppers and landlords on technological progress in Northeast Brazil. The approach parallels Badhuri's (1973) contribution on the study of rural backwardness in West Bengal. Unlike Badhuri's model, however, risk is explicitly taken into account, and the case of land- and labor-augmenting technological progress is considered.

The results indicate that under certain conditions sharecroppers and landlords are both likely to oppose technological progress. The sharecroppers would oppose it because, although producing higher expected incomes, it would also be associated with an increase in riskiness of the farming operation. More interestingly, the landlords would oppose it too because of the consequent weakening of the dependence bond with their sharecroppers. The landlords benefit from the dependence relationship through usury and monopolistic marketing.

The analysis also demonstrates that different types of innovations are likely to elicit opposite responses from the parties involved. Land-augmenting innovations would be favored by sharecroppers in a regime of increasing interest rates but support by landlords would be unlikely. Labor-augmenting innovations, on the other hand, would be accepted by sharecroppers only at sufficiently low interest rates but would generally be promoted by landlords. In light of this, labor-augmenting technological progress would seem to be

The author is grateful to Bela Balassa, Alain de Janvry, and Peter Hazell for helpful comments. The views expressed in this paper are his own and do not necessarily reflect those of IBRD.

the most likely to be accepted without major conflicts between landlords and tenants in the Northeast.

SHARECROPPERS IN THE SERTÃO

Production System

Most of the interior of Northeast Brazil consists of a harsh, semiarid, drought-prone backland called the Sertão. The agriculture characteristic of this part of the world is extremely primitive and is based on a mixture of swidden (slash and burn) and long-term fallow ecotypes. (For the definition of agricultural ecotypes, see Wolf, 1966.) The backwardness of the Sertão farmer is certainly dramatic; not only modern inputs but even the most primitive forms of plowing are virtually unknown. The basic tool is the hoe, and only a minimal part of the farm population has experience with anything more sophisticated than a manual planting tool.

Farmer resistance to technological change has been explained with different arguments—from a regressive value system to active pragmatism and response to risk (see Johnson, 1970, 1971). Although these arguments may help understanding of why small farmers do not innovate, they do not provide any justification as to why large landowners—who are much less risk prone, relatively well educated, and well endowed with money—do not invest in technological change and do not try to diffuse it among the scores of peasants working for them.

In the Sertão the most widespread form of land tenure arrangement is sharecropping, whereby medium and large landowners divide their land among a number of (generally) landless peasants in exchange for a share of the crop. In the large *fazendas,* part of the land regulated by this form of contract is incorporated in a sizable unit run by the owner and/or by an administrator. In this case, sharecropping is much closer to a form of labor contract than to a land-rental contract.

Perpetual Indebtedness and Economic Dependence

The sharecropper is typically involved in the production of one cash crop (cotton) and a number of subsistence crops (corn, beans, and manioc). In its most diffused form the sharecropping contract is very simple, since it involves only the transfer of a share of the net harvest (total harvest minus seed for planting next year) of the cash crop to the landlord. Inputs are generally not provided, and the sharecropper is free to plant as much subsistence crop as he wishes. Production of the cash crop, however, must be kept at a level considered satisfactory by the landlord. In the areas where commercial agriculture is more developed, there is often an active market for staples, and a share of the subsistence crop (though in general smaller than the share of the cash crop) is also due to the landlord.

The sharecropper can be said to depend on the landlord in three main ways. First, the landlord obviously controls use of land. Also, housing is often provided with the land and, in large *fazendas,* a primitive infrastructure. Second, the landlord is the sole intermediary between himself and the market. Third, he is the main source of credit when cash or food is needed.

Granting of credit and purchase of the harvest are closely related. Two or three months before harvest, when prices are at their highest and the sharecroppers are out of cash and food, the owner will buy crops green, at approximately one half their market value when ripe. Alternatively, he will sell staples on credit at current market prices, which may be several times greater than after harvest.

Because of the high interest rates; the comparatively low prices for the product; and the wild oscillation of production due to pests, diseases, and periodic droughts, the sharecropper is often caught in a relation of perpetual indebtedness and dependence on the landlord. This relation is, however, not forced upon the farmers but accepted by them as a sort of social form of insurance against the exteme risks of the environment. As Johnson (1970) points out: "In short, the landlord is expected by the *moradores* to give more than merely rights to land. Far from hating the existence of a company store or the purchase of green crops at cut-rate prices, the workers regard these practices as the only alternative to great potential suffering. . . . In the face of great uncertainties, a worker has no ties that can assure him as much protection as a firm tie to the landlord can."

Technological Progress

To a casual observer it may seem that technology in the Sertão is so primitive that it would take only a minimum of initiative by workers or landowners to produce almost immediate increases in productivity. Indeed, local extension agents have assisted in introducing plowing, proper spacing in planting, and use of selected seed on several farms. However, these farms are a small minority; innovations tend to be limited to the few, medium-sized farms directly reached by the extension programs.

In the large units resistance to technological innovation seems to come almost equally from the workers as from the landowners. The workers resist innovations mostly by refusing to apply additional labor to techniques with uncertain results, such as plowing or spacing. The owners simply fail to reinvest their profits in improvement of their property, the capital endowment of which is kept at the minimum level compatible with the going technology.

The main explanation of sharecropper resistance to change is undoubtedly risk. Most of the new techniques (including use of modern inputs such as fertilizers, selected seed, etc.) are riskier in the sense that the higher expected returns are counterbalanced by higher variability. Furthermore, without access to modern capital markets, the workers are in no condition to invest in innovations. The lack of initiative of the landlords in this respect is more difficult to explain. In questioning large landowners, one finds that they are involved in other more "dynamic" activities such as commerce and construction, and that they consider their farms more as an asset than as a productive unit. More traditional landlords tend to consider the farm as a static rent-generating asset, and they would rather invest in beautification than in technological improvements.

The mathematical model that follows will try to capture the two elements of resistance to technological innovation: (1) risk aversion and the progressive financial weakness of the sharecroppers and (2) interest of the landlord in maintaining the status quo. The hypothesis underlying the model will be that the relation of economic dependence between landlords and sharecroppers is exploitative and tends to delay economic progress.

THE MODEL

The Sharecropper's Side

Some of the symbols used are the following:

z_t = "net worth" of available stock of output of the sharecropper
c_t = value of consumption of the sharecropper
x_t = value of production
r = share of production retained by the sharecropper
i = interest rate "imposed" on credit by the landlord
L = labor endowment of the sharecropper
ω = opportunity cost (e.g., wage rate) of the sharecropper's labor

All the variables without the time subscript are assumed to be nonstochastic and independent of time. Production is regulated by the following stochastic production function:

$$x_t = \epsilon_t F(L) \qquad (8.1)$$

where ϵ_t is the realization at time t of a random variable ϵ (which may be taken to represent yield), with mean μ and variance σ^2. As Badhuri (1973) suggests, rather than depending on income, consumption may be assumed to depend on the balance of the available output in each period:

$$c_t = bz_t + b_0 \qquad (8.2)$$

Since land and farm capital are predetermined (fixed by the landlord and/or institutional constraints) and modern inputs are assumed to be available in the base situation, the specified production function contains only the labor variable. The sharecropper maximizes his expected utility by setting a value for this variable, say L^*. The available stock of output can then be defined as

$$z_t = r\epsilon_t F(L^*) - (1+i)(c_{t-1} - z_{t-1}) + \omega(\bar{L} - L^*) \tag{8.3}$$

Substituting equation (8.2) into equation (8.3), taking expectations on both sides, and solving the difference equation $Ez_t = r\mu F(L^*) + \omega\ (\bar{L} - L^*) + (1+i)\ (1+b),\ Ez_{t-1} - (1+i)b_0$:

$$Ez_t = \bar{E}z + (Ez_0 - \bar{E}z)\,[(1+i)(1-b)]^t \tag{8.4}$$

where $\bar{E}z$ indicates the asymptotic value of expected net worth and is equal to

$$\bar{E}z = \frac{r\mu F(L^*) + \omega\bar{L} - (1+i)b_0 - \omega L^*}{1-(1+i)(1-b)} \tag{8.5}$$

and Ez_0 indicates the expected value of initial net worth. The condition for convergence of equation (8.5) is $i < b/(1-b)$. If the rate of interest imposed by the landlord respects this limit, there will be a steady state for farmer expected output balance. The steady state will be characterized by a constant level of indebtedness, provided that $rF(L^*) + \omega\bar{L} < (1+i)b_0 + \omega L^*$. In this case,

$$\begin{aligned} \bar{E}D &= \bar{E}c - \bar{E}z = (b-1)\,\bar{E}z + b_0 \\ &= (b-1)\,\frac{\mu r F(L^*) + \omega\bar{L} - (1+i)b_0 - \omega L^*}{1-(1+i)(1-b)} + b_0 \end{aligned} \tag{8.6}$$

where $\bar{E}D$ indicates the asymptotic value of expected indebtedness. Thus the expected long-run level of sharecropper debt will be positive if

$$r\mu F(L^*) + \omega(\bar{L} - L^*) < (1+i)b_0 \tag{8.7}$$

or, in words, if the average amount of produce retained by the sharecropper, plus the amount he earns from wage labor, is less than the value of a loan to satisfy the minimum consumption needs. Notice that in the opposite case, i.e., when $r\mu F(L^*) + \omega\bar{L} - \omega L^* \geq (1+i)b_0$, the sharecropper will not borrow. I assume that the sharecropper is not able to lend at a rate of interest $\geq i$. Thus if $z\mu F + \omega(\bar{L} - L^*) > (1+i)b_0$, the farmer will be able to consume above the minimum level b_0 without borrowing.

Notice also that the limit established for i is most likely to be respected if propensity to consume is high. Since $\lim_{b \to 1}$, $b/(1-b) = \infty$, it is clear that as b approaches 1, the constraint on i is lifted and any value of the interest rate becomes compatible with convergence and a positive value of steady-state expected net worth. For $b = 0.5$, only values of i less than 30 percent would ensure convergence.

If $i > (1-b)/(1-b)$, equation (8.4) will diverge. In this case expected indebtedness will be

$$ED_t = \bar{E}D + (ED_0 - \bar{E}D)\,[(1+i)(1-b)]^t \quad \textbf{(8.8)}$$

where $\bar{E}D$ is formally described again by equation (8.6) and ED_0 stands for expected value of initial indebtedness. In particular, $r\mu F(L^*) + \omega(\bar{L} - L^*) < (1+i)b_0$ and the term $ED_0 - \bar{E}D$ will be necessarily positive since

$$ED_0 - \bar{E}D = ED_0 + (1-b)\,\frac{(1+i)b_0 - r\mu F(L^*) - \omega(L - L^*)}{(1+i)(1-b) - 1} + b_0 > 0 \quad \textbf{(8.9)}$$

so that sharecropper debts will grow without limit. The same result will hold if initial debts plus minimum consumption $ED_0 + b_0$ exceed the asymptotic debt; i.e., $ED_0 + b_0 > \bar{E}D$.

Both equations (8.8) and (8.9) make clear the role that an increment in production or consumption could play. Technological progress could increase $r\mu F(L^*)$ over $(1+i)b_0$, liberating the farmer from his dependence on the landlord, and so could suppression of consumption related to ceremonial or other social requirements.

The Problem of Uncertainty

We obtain the expression for variance of output balance as defined in equation (8.3). Applying the variance operator V to both sides of this equation and solving the resulting difference equation yields

$$V(z_t) = \bar{V}z + (Vz_0 - \bar{V}z)\,\{1 - [(1+i)(1-b)]^{2t}\} \quad \textbf{(8.10)}$$

where $\bar{V}z$ indicates the asymptotic value of variance $V(z_t)$ and is equal to

$$\bar{V}z = \frac{r^2\sigma^2[F(L^*)]^2}{1 - (1+i)^2(1-b)^2} \quad \textbf{(8.11)}$$

and Vz_0 is the variance of output balance at time zero. As readily seen by inspection of equations (8.10) and (8.11), the conditions for convergence of variance are the same as for the expected value of z_t.

Assume now that sharecropper behavior conforms to a utility function linear in the expected value and the standard deviation of output balance. In the absence of other restrictions and with a sufficiently long time horizon, maximizing a linear function of the two statistics defined as in equations (8.4) and (8.10) is equivalent to maximizing the same function of their asymptotic terms.

Algebraically, we can state the problem as follows:

$$\max_L \bar{E}U = \bar{E}z - \Phi(\bar{V}z)^{1/2}$$

$$= rF(\lambda L)\left(\frac{\mu}{1-(1+i)(1-b)} - \Phi \frac{\sigma}{\sqrt{1-(1+i)^2(1-b)^2}}\right) - \frac{\omega L + (1+i)b_0}{1-(1+i)(1-b)} \quad (8.12)$$

where $\Phi > 0$ is a risk-aversion coefficient and λ is a multiplier for labor-augmenting technical progress.

Assuming the size of the plot as given, first-order conditions for the maximization of equation (8.12) are

$$\frac{d\bar{E}U}{dL} = rF'\left(\frac{\mu}{1-(1+i)(1-b)} - \Phi \frac{\sigma}{\sqrt{1-(1+i)^2(1-b)^2}}\right) - \frac{\omega}{1-(1+i)(1-b)} = 0 \quad (8.13)$$

Consider first the case of neutral (or in the context, land-augmenting) technical progress. Such a case can be represented by an increase of the expected value of yield per acre μ. Differentiating equation (8.13) with respect to μ, we obtain

$$\frac{rF'}{1-R} - \frac{r\Phi F'}{\sqrt{1-R^2}} \frac{d\sigma}{d\mu} + rF''\left(\frac{\mu}{1-R} - \frac{\Phi\sigma}{\sqrt{1-R^2}}\right) \cdot \frac{dL}{d\mu} = 0 \quad (8.14)$$

where $R = (1+i)(1-b)$.

For technical progress to be advantageous to the sharecropper, we must have

$$\frac{d\bar{E}U}{d\mu} = rF(L)\left(\frac{1}{1-R} - \frac{\Phi}{\sqrt{1-R^2}}\frac{d\sigma}{d\mu}\right)$$

$$+ rF'\left(\frac{\mu}{1-R} - \frac{\Phi}{\sqrt{1-R^2}}\right)\frac{dL}{d\mu} - \frac{\omega}{1-R}\frac{dL}{d\mu} \geq 0 \qquad (8.15)$$

Substituting the value of $dL/d\mu$ given by equation (8.14) into equation (8.15) and solving for $d\sigma/d\mu$, we obtain

$$\frac{d\sigma}{d\mu} \leq \frac{\sqrt{1-(1+i)^2(1-b)^2}}{\Phi[1-(1+i)(1-b)]} = G(i,b) \qquad (8.16)$$

Thus land-augmenting technological progress will be beneficial only if the associated increase in risk (if any), as measured by the yield standard deviation, is less than a function of the interest rate and the propensity to consume.

From equation (8.16) it is clear that $G(i, b)$ is an increasing function of i and a decreasing function of b. Differentiating both sides of equation (8.16), we obtain

$$\frac{\partial G(i,b)}{\partial i} = \frac{(1-R^2)^{1/2}(1-b)-(1-R^2)^{-1/2}(1-b)^2(1+i)(1-R)}{(1-R)^2} \qquad (8.17)$$

$$\frac{\partial G(i,b)}{\partial b} = \frac{(1-R^2)^{-1/2}(1-b)(1+i)^2(1-R)-(1-R^2)^{1/2}(1+i)}{(1-R)^2} \qquad (8.18)$$

From equations (8.17) and (8.18) we can see that $\partial G/\partial i > 0$ and $\partial G/\partial b < 0$ if the condition of convergence $(1+i)(1-b) < 1$ is respected. Reverse inequalities will hold in the divergence case. This result may seem paradoxical. Inspection of equation (8.12), however, provides a simple explanation. From this formula one can readily see that introduction of credit in the model has a different impact on the marginal utilities of mean and standard deviation of yield (or sharecropper revenue). The multipliers of μ and σ in the term in parentheses imply that increases in the interest rate would, within the convergence limits, increase both the risk term and the mean term. For a given Φ, however, the increases in the weight of the mean Φ would be greater than the increases in the weight of the standard deviation.

This result is entirely a consequence of the debt mechanism described: as the cost of his credit increases, the sharecropper's risk aversion increases less proportionately than his need. In terms of the land-augmenting innovation examined, one can also say that in equilibrium the need for repayment affects wealth statistics differently. Thus a unit increase in mean yield would make the farmer richer by $1/[1-(1+i)(1-b)]$ in terms of expected wealth, while the risk measure would increase only by $1/\sqrt{1-(1+i)^2(1-b)^2}$.

From this summary analysis we can conclude that land-augmenting technological progress can be made more attractive to the sharecropper by a policy of high interest rates when the same policy does not result in a condition of progressive indebtedness.

Consider now the case of labor-augmenting technological progress. Differentiating both sides of equation (8.13) with respect to λ, we obtain

$$r\left(\frac{\mu}{1-R} - \Phi\frac{\sigma}{\sqrt{1-R^2}}\right)(F' + F''L)\frac{dL}{d\lambda} - \frac{r\Phi F'}{\sqrt{1-R^2}}\frac{d\sigma}{d\lambda} = 0 \quad \textbf{(8.19)}$$

where we have assumed $\lambda = 1$ in the initial position. For the innovation to be adopted, it must be

$$\frac{d\bar{E}U}{d\lambda} = \left[rF'\left(\frac{\mu}{1-R} - \Phi\frac{\sigma}{\sqrt{1-R^2}}\right) - \frac{\omega}{1-R}\right]\frac{dL}{d\lambda} - \frac{\omega}{1-R}L$$
$$- \frac{r\Phi F}{\sqrt{1-R^2}}\frac{d\sigma}{d\lambda} \geq 0 \quad \textbf{(8.20)}$$

Solving equation (8.20) after substituting the value of $dL/d\lambda$ obtained from equation (8.19) yields, as the "acceptance" condition,

$$\frac{d\sigma}{d\lambda} \leq \frac{\sqrt{1-(1+i)^2(1-b)^2}}{\Phi[1-(1+i)(1-b)]}\left[\frac{(F' + F''L)\omega L}{rF'\alpha - \dfrac{\omega}{1-R}}\right] \quad \textbf{(8.21)}$$

where $\alpha = [\mu/(1-R)] - \Phi(\sigma/\sqrt{1-R^2})$. If we take $\sigma = 0$ (i.e., we consider the limit of the variation over the present value of variance), expression (8.21) simplifies to

$$\frac{d\sigma}{d\lambda} \leq \sqrt{1-(1+i)^2(1-b)^2}\left[\frac{\omega L(F' + F''L)}{\Phi(rF' - \omega)}\right] = M(i,b)/\Phi \quad \textbf{(8.22)}$$

Taking the partial derivatives of M with respect to i and b, we obtain:

$$\frac{\partial M}{\partial i} = -\left[\frac{\omega L(F' + F''L)}{rF' - \omega}\right]\frac{(1+i)(1-b)^2}{\sqrt{1-(1+i)^2(1-b)^2}} \quad \textbf{(8.23)}$$

$$\frac{\partial M}{\partial b} = \left[\frac{\omega L(F' + F''L)}{rF' - \omega}\right]\frac{(1+i)^2(1-b)}{\sqrt{1-(1+i)^2(1-b)^2}} \quad \textbf{(8.24)}$$

Thus for $F' + F''L > 0$, M is a decreasing function of the interest rate and an increasing function of the propensity to consume. In this case, higher rates of interest will discourage farmer adoption of new techniques and, somewhat paradoxically, so will low propensities to consume. This result runs counter to the one obtained for land-augmenting innovation. It is due to the fact that, in the model, increases in labor productivity per se do not affect the mean of land productivity, but only its variance. Analogous results are obtained if one assumes that labor-productivity increases are associated with increases in the coefficient of variation of yield.

From the analysis above it follows that land-augmenting and labor-augmenting innovations may play quite a different role in the system of dependence described. The result may be taken qualitatively to hold for the so-called "landesque" and "laboresque" innovations described by Sen (1959) and the "biological" and "mechanical" innovations described by Hayami and Ruttan (1971).

Land-augmenting technological progress would be acceptable to risk-averse sharecroppers even under a regime of high financial dependence from the landlords. Provided that the increasing interest rates do not cause bankruptcy, adoption rates in this case would indeed be higher as interest rates rose. Labor-augmenting technological progress, on the other hand, will be attractive only if interest rates are sufficiently low to make the undertaking profitable in the face of increasing risks.

The Landowner's Side

Consider now the point of view of the landowner. The expected income of the landlord from a single sharecropper, under the assumption that both the contract share and the plot size are given, can be defined as

$$Ey_t = (1-r)\mu pF(L) + r\mu(p-1)F(L) + i(Ec_{t-1} - Ez_{t-1}) \quad \textbf{(8.25)}$$

where p is the ratio between the price the landlord obtains in the free market and the price he pays to the sharecropper, and i is the net interest he is able to charge on the loans. Simplifying and substituting equations (8.2) and (8.4) into equation (8.25) yields

$$Ey_t = \bar{E}y + (Ey_0 - \bar{E}y)\,[(1+i)(1-b)]^{t-1} \quad \textbf{(8.26)}$$

$$\bar{E}y = (p-r)\mu F(L) - i(1-b)\,\bar{E}z + ib_0 \quad \textbf{(8.27)}$$

Assume first $i < b/(1-b)$. In this case equilibrium income converges in mean to $\bar{E}y$. If the landlord is risk neutral, he will accept technical progress to the extent that the corresponding increase in production in-

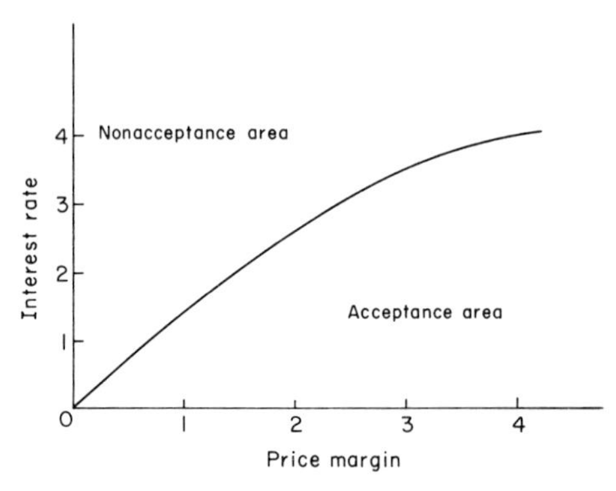

Fig. 8.1. Relation between the interest rate and the price margin.

creases his expected income. In the case of land-augmenting technological progress,

$$d\bar{E}y/d\mu = (p-r)F(L) - i(i-b)\,(d\bar{E}z/d\mu) \geq 0 \tag{8.28}$$

Substituting the explicit expression for $d\bar{E}z/d\mu$ obtained from equation (8.5) yields the two alternative conditions in (8.29). For simplicity we assume $dL/d\mu = 0$, i.e., from expression (8.16) $d\sigma/d\mu = (1/\Phi)\,[\sqrt{1-R^2}\,/\,(1-R)]$.

$$i < \frac{b}{1-b}\left[\frac{p-r}{1+(p-r)}\right] \quad \text{or} \quad (p-r) > \frac{(1-b)i}{b-(1-b)i} \tag{8.29}$$

Thus only if the interest rate is sufficiently low and/or the price margins exacted from the sharecropper are high enough will the landowner be in favor of land-augmenting technical progress. The relation between p and i is demonstrated graphically in Figure 8.1. The curve is the limit between the acceptance and nonacceptance areas and is drawn for a value of the propensity to consume equal to 0.8. As the figure demonstrates, for values of interest rates of the order of magnitude observed in practice, landlords are not likely to favor neutral technological progress unless they can renegotiate some of the terms of the contract. Even if they can, upper bounds on the new terms are likely to exist. Thus it is reasonable to assume that the share of the crop cannot go beyond 0.60 ÷ 0.65, since legislation is already requiring it to be not greater than 0.5. Similarly, p cannot be much greater than 1, or the sharecroppers would be motivated to divert increasing quantities of their produce to other buyers.

But what if $i > b/(1-b)$? This is a case either of very high interest rates or of very strong requirements on the part of the sharecropper to keep a large stock of output unconsumed. The asymptotic equilibrium associated with $\bar{E}z$ is now unstable, and any increase in productivity will move the farmer further on his exponential growth path—expression (8.4)—and move the landowner to the corresponding path described by equation (8.26). Differentiating this last equation with respect to μ yields

$$\frac{dEy_t}{d\mu} = (p-r)\,\frac{i(1-b)}{1-(1+i)(1-b)}\left\{1-[(1+i)(1-b)]^{t-1}\right\} \qquad (8.30)$$

This derivative has to be ≥ 0 for an increase in μ to be profitable to the landlord. Simplifying and solving for p,

$$p = r + \frac{i(1-b)[(1+i)(1-b)]^{t-1}-1}{(1+i)(1-b)-1} \qquad (8.31)$$

Therefore, in the case considered, any profit the landlord may have from technological progress will vanish very rapidly as t grows, since the autonomous increase in production will progressively free the sharecropper from his dependence bond.

In the case of labor-augmenting technological progress the results are somewhat similar. Here, to avoid triviality we must consider explicitly the variation in labor input of the sharecropper as obtained in equation (8.19). Assuming $d\sigma/d\lambda = 1$, the expression for the variation of asymptotic income of the landlord is

$$\frac{d\bar{E}y}{d\lambda} = \frac{[(p-r)\mu F'(1-R)-i(1-b)(r\mu F'-\omega)]r\Phi F'}{\mu\sqrt{1-R^2}-\sigma(1-R)} + \frac{\omega}{1-R}L \qquad (8.32)$$

The value of this expression must exceed zero for the innovation to be attractive to the landlord. A sufficient condition for this is that the first term in equation (8.32) be greater than zero; i.e.,

$$p \geq [i(1-b)\omega^* + br]/[1-(1+i)(1-b)] \qquad (8.33)$$

where $\omega^* = \omega/\mu F'$ is given by the condition in equation (8.13). For plausible values of the parameters involved and reasonably high rates of interest, this condition is less restrictive than the condition found for land-augmenting technological progress. For example, for $r=0.5$, $0=1$, $\sigma=\mu=1$, $b=0.8$, and $i=1.5$, p could be as low as 1.1 with technological progress still benefiting the landlord. This lower bound for p increases very rapidly, however, with increases in i. For example, for the same values of the other parameters and $i = 3$, p should be at least 2.90 for the innovation to be attractive.

Thus labor-augmenting technological progress is more likely to be attractive to the landlord and he could be motivated to decrease the rates of interest imposed on loans to make it more attractive to the sharecropper.

CONCLUSIONS

The results presented suggest a number of qualitative conclusions. First, the willingness of the sharecropper to undertake risky technological progress is bound to be low because of his weak economic position. Second, the risks as well as this weakness are compounded by the high interest rates imposed by the landlord.

From the other side, the landlord may have incentive for productivity increases only if the consequent increase in income does not weaken the bond of dependence that links the sharecropper to him. If this occurs, the landlord may be able to achieve better economic results by combining usury and marketing with his land-renting activity.

In a situation characterized by periodic drought and little or no financial solvency of workers (as in Northeast Brazil), usury may generate a vicious circle. From one side the landlord will tend to impose high interest rates that will weaken the sharecropper's economic position and may decrease willingness to undertake risks in production. The same high interest rates, on the other hand, may produce a situation of dependence, exploitation, and rent that will make it unprofitable for the landlord to accept technological progress.

The analysis suggests also that a conflict of interests between landlord and sharecropper is more likely to arise in the case of land than with labor-augmenting innovations. Land-augmenting innovations, although presumably favored by workers under a regime of increasing interest rates, would be likely to be opposed by landlords for the reasons suggested above. Landlords would have, on the contrary, a definite incentive for promoting labor-augmenting innovations by decreasing the rates of interest or providing alternative incentives to the sharecroppers.

This conclusion is particularly relevant for the Northeast, since the agriculture of this region is very primitive—essentially "a man and a hoe" affair. Therefore, the possibility of both types of innovation exists and, interestingly, "laboresque" innovations need not be confined to extreme labor-saving forms of modernization, as they include a whole range of intermediate technologies such as plowing, using draft animals in traditional operations, clearing by mechanical means, removing stumps, etc.

There is considerable evidence that "biochemical" innovations in the

Northeast are associated with small increments in expected yields and significant increases in production risks. This is especially true for fertilization and "selected" varieties, but it is also true for the use of some pesticides and weed-control chemicals. The adoption of slightly more modern agricultural practices of the "mechanical" type, on the other hand, appears to have a considerable effect on labor productivity without sensible increases in risk. For example, preliminary results of the IBRD-SUDENE survey and a 1975 IBRD study show that labor productivity of farmers using animal power is roughly twice as much as the productivity of other farmers. The increase in risk associated with such an increase in labor productivity seems to be insignificant.

COMMENT / *Alain de Janvry*

The model used by Scandizzo is an extension of Badhuri's (1973) work on sharecropping. In Badhuri's model, sharecroppers are related to the landlord through payment of a rent in kind, permanent indebtedness and usurious loans, and monopolistic control by the landlord of the farmer's access to the market. To this model Scandizzo adds agronomic risk in production and laborsaving and/or landsaving technological changes.

The fundamental question asked is, How do sharecroppers and landlords—in a land tenure system characterized by rent in kind, usury, and monopolistic market control—behave toward the adoption of risky landsaving and laborsaving technological changes? The question is motivated by the observed extreme technological backwardness of Northeast Brazil in spite of the presumed availability of new technologies for adoption. It is reminiscent of "the *latifundio* puzzle of Professor Schultz" (Feder, 1967), who states, "One would expect that farmers who operate large enterprises would actively search for new agricultural factors. . . . Why they have not done better on this score is a puzzle" (Schultz, 1964).

To answer this broad and difficult question, Scandizzo uses a highly simplified and abstract model of choice. The basic answers derived from the model are that (1) technological change is resisted by sharecroppers due to risk and by landlords because it decreases the gains from usury and (2) biochemicals (landsaving technology) are favored by sharecroppers, especially if interest rates are high (!), but they are opposed by landlords. By contrast, mechanical innovations (laborsaving technology) are acceptable to sharecroppers if interest rates are low, and they are also favored by landlords. While highly mysterious to interpret, the implication is that mechanical innovations are likely to diffuse more easily than biochemicals under the prevailing land tenure system.

In my view the model has contradictions in its formulations and severe limitations in ability to answer the question raised. The formulation of the model contradicts several of the structural characteristics stated as prevailing in the Northeast:

1. It is observed that successful introduction of new technologies has occurred on farms assisted by the extension service. If this is indeed the case, the limiting factor to acceptance of technology would seem to be the lack of information available to the mass of farmers. In this case considerations for risk (sharecroppers) and for the weakening of usury bonds (landlords) would seem to be irrelevant.
2. It is stated that "without access to modern capital markets, the workers are in no condition to invest in innovations." Again, if this is indeed the case, lack of access to financial institutions is the limiting factor for sharecroppers, and considerations for risk are only of secondary importance.
3. It is observed that land is held as a static rent-generating asset in a diversified portfolio (which also includes nonagricultural investments) rather than as a factor of production. This is indeed an observation that characterizes land tenure in almost all Latin America. As we now know, this is not due to the lack of entrepreneurial drives and talents among landowners but is due to the simple fact that modern farming is unprofitable under the existing terms of trade for agriculture. It is this unprofitability that makes technological backwardness and absenteeism economically optimum options (see e.g., Obschatko and de Janvry, 1972). Yet the question of profitability of the presumed available new technologies is not even raised, in spite of the above observation.

Scandizzo says, "In my model the share of the crop r and the marketing margins p of the landlord *are not* assumed to be fixed. Results are simply obtained that relate the share of the crop, the marketing margins, and the increase in risk σ with the effects of technical change. Thus all the formulas obtained in the part concerning the landlords trace the boundaries that technological progress has to respect in these three dimensions (r, p, and σ)" (personal communication to the editors).

A severe limitation of the model to answer the question raised results from considering the sharing of the crop between landlord $(1-r)$ and sharecropper (r) as a fixed, exogenous variable unrelated to the level of product generated. This is clearly untenable since variation of this share is precisely the dominant mechanism of surplus extraction under sharecropping, and it is thus not exogenous relative to surplus generation. The assumption of a fixed r presumes the answer of the model—landlord antagonism to technological change. Indeed, what is exogenous is the subsistence level of the sharecropper household—defined here by the minimum consumption level b_0 or by the opportunity cost ωL on the labor market (whichever is the higher); the landlords adjust r to siphon out whatever surplus the sharecropper produces above this level. In this case, landlords are perfectly happy with technological change as long as it is profitable; i.e., as long as it increases the surplus they extract from the farmers. This, of course, does not mean that usury is not an important aspect of semifeudal *fazendas*. Usury permits the landlords to tie sharecroppers to the land. It is thus not so much a direct mechanism of surplus extraction as it is a means of alienating the farmers from capturing their own opportunity costs on the labor market. This lowering of the value of labor to minimum subsistence requirements b_0 permits r to increase and to maximize surplus extraction via payment of a rent in kind.

Fixing of r also presumes the behavior of sharecroppers and landlords toward the bias in technological change: if they cannot renegotiate r, landlords prefer mechanization that is not yield increasing (and thus maintains the status quo of exploitation via usury) to yield-increasing (and thus sharecropper-liberating) biochemicals. The mysterious inverse relationships between adoption of biochemicals under high interest rates and of mechanicals under low rates probably result from the difference between coercive and incentive effects of interest rates for each type of technology. Because biochemicals are yield increasing, farmers can be forced to modernize since they have to satisfy minimum subsistence requirements; because mechanicals are laborsaving, they will only be adopted if there is a monetary incentive to do so.

Returning to the central question of the relationships between technological change (development of the forces of production) and land tenure (the social relations of production), Badhuri openly acknowledges that this is a problem that pertains to the domain of historical materialism. A neoclassical formulation of farm management under sharecropping is useful to conceptualize and eventually to measure surplus transfers between sharecroppers and landlords. But it is incapable of explaining the evolution of the social relationships concerned with the development of the mode of production when this is the central question in regard to sharecropping.

What need to be identified are the conditions under which sharecropping tends to prevail as a land-tenure arrangement and the conditions under which sharecropping tends to disappear to give way to full-fledged capitalistic commercial agriculture.

Clearly, sharecropping (rent in labor services and rent in kind) is only a step in the development of capitalism that tends to prevail under (1) primitive development of the product market ("natural" manorial economy in Western Europe from the eleventh to the eighteenth century) or lack of easy access to local markets for the farmers, who then market their commodities via the landlords (the Latin American export-oriented *hacienda*); (2) labor scarcity and primitive development of labor markets, where tying the farmers to the land through debts or superstructural obligations (bonded labor) is essential to secure cheap labor for the *latifundio;* and (3) low profitability of the development of productive forces whereby imposition of usurious rents is more profitable for landlords than application of technology.

Latin American history clearly reveals an evolution from sharecropping (rent in labor services and in kind) to capitalistic farming (rent in cash or owner-operated enterprises) in relation to (1) the rise of surplus labor—particularly notable since the 1950s and for this reason an important determinant of land-reform programs in the 1960s—and (2) the profitability of farming. In Chile, "junkerization" of the *latifundio* has occurred in discontinuous jumps that are directly related to the successive periods of opening of profitable markets for grain export. In Argentina, *la obligación* (rent in labor services) disappeared in the early 1800s with the opening of a lucrative market for export of cattle to England and is now found only in remote areas of the northwest. In Brazil, sharecropping is virtually nonexistent in the Rio Grande Do Sul, with its profitable cattle (and recently soybean) activities, while it still dominates in the Northeast.

What are, ultimately, the major blockages to adoption of technology in Latin American agriculture? Three major causes have been advanced: (1) the backward land tenure system dominated by the *latifundio* and characterized by absenteeism and exploitative domination over sharecroppers (the CIDA "structuralist" thesis) (see, e.g., Barraclough, 1973); (2) the unavailability of modern technology due to lack of research or information; (3) the unprofitability of adoption of technology due to high-priced industrial inputs (import substitution industrial policies) and low food prices (overvaluation of exchange rates, price controls as part of antiinflationary policies, and "cheap food policies") (Schultz, 1968) in general. Lack of profitability as a determinant of stagnation has been denounced by "monetarists" (e.g., Schultz and Johnson), thus proclaiming the need to eliminate market distortions and return to the free market. It has also been denounced by the defendants of the "unequal development" theory (Baran, Frank, and Amin). For the latter, the origin of cheap food policies is found in the logic of cheap labor associated with the process of accumulation in a socially disarticulated industrial structure (see Ch. 11).

For "monetarists" and "unequal development" theoreticians, the *premium mobile* for the adoption of agricultural technology under a capitalistic system is profitability. Once profitability obtains, the other two bottlenecks disappear: it creates an effective demand for new technological research and information and thus induces supply by research institutions, private agribusiness firms, and extension services; and the consequent development of capitalism transforms social relationships of production, including the land tenure system and the forms of payment to sharecroppers.

Permanence of traditional, absentee, exploitative farming systems is simply rational under conditions of unfavorable terms of trade, and this is what Scandizzo rediscovers. But in no way should the archaic land tenure system be taken as a determinant of technological backwardness, an implication Scandizzo draws by not raising the question of profitability and thus confusing fact and essence. The essence is unprofitability, which in turn finds its logic in the contradictory laws of accumulation in a disarticulated structure. The facts are the permanence of archaic land tenure patterns and technological stagnation.

III

Technology, Rural Development, and Welfare

CARLOS A. ZULBERTI
KENNETH G. SWANBERG
HUBERT G. ZANDSTRA

9

Technology Adaptation in a Colombian Rural Development Project

The purpose of this chapter is to define the role of economic analysis in the design of new technology for small farmers. In establishing this definition, the authors refer to some of the experiences acquired in a specific rural development project in Colombia during more than four years of work with small farmers. The information to be reviewed was drawn from the Eastern Cundinamarca Project (generally known as the Cáqueza Project), one of the twenty-one rural development projects the Colombian Agricultural Institute has initiated.

We assume that the new technology in question should satisfy, as a minimum, two important objectives: the increase of small-farmer incomes and the increase in total production for the community. We also believe that the real value of a new technology for small farmers can only be measured by their adoption of it. No technology is good for the small farmer if he does not or cannot adopt it.

ADJUSTING NEW TECHNOLOGIES TO THE REGION

The Cáqueza Project began in 1971 with establishment of more than thirty experiments for adjusting the existing new technologies in corn and potato production to the specific conditions of the region. Priority was given to these two crops because they were most frequently encountered in the area. Sixty-three percent of the cultivated hectares under project influence were

The authors are members of the Interdisciplinary Rural Development Research group of the ICA-IDRC Cooperative Agreement. The opinions expressed in this document are those of the authors and do not necessarily represent those of the sponsoring institutions.

seeded to corn, primarily in association with a legume; 27 percent were seeded to potatoes, either alone or with legumes. After the first year of experimentation, which was carried out in small plots in the farmers' fields using traditional cultivation practices, the project was able to identify new technologies for corn and potatoes that increased production and income without causing a corresponding decrease in production of the associated crops. The corn recommendation (see Table 9.1) increased production 202 percent and increased net income 253 percent, while the recommendation for potatoes (see Table 9.2) increased production 51 percent and net income 30 percent.

Information on these new technologies was widely communicated to farmers in the area. At the same time, a supervised credit program was made available by the government agricultural credit bank (Caja Agraria) to finance the necessary material inputs (fertilizers, seeds, and pesticides). To receive this credit, the farmer had to prepare a production plan according to specifications of ICA agronomists, wherein the farmer had to agree to apply the recommended package through the crop-production cycle.

Table 9.1. Traditional and new technology for corn production

Production information	Traditional technology	New technology
Variety or hybrid	Criolla	H208-255-302[*]
Seeding density		
Rows/ha	90	100
Plants/100 meters of row	90	125
Plants/hole	3	3
Fertilization		
10-30-10	0	200 kg
Urea: first application (at seeding)	0	25 kg
Urea: second application (40-50 days after seeding)	0	125 kg
Pest control		
Earworm (*Heliothis* spp.)	No control	Control
Armyworm (*Spodoptora rugiperda*)	No control	Control
Cutworm (*Agrotis* spp.)		
Production		
Average yield	907 kg/ha	2740 kg/ha
Standard deviation	660 kg/ha	1170 kg/ha
Net earnings	US$58/ha	US$205/ha

[*]The recommended hybrid varies according to altitude.

Table 9.2. Traditional and new technology for potato production

Production information	Traditional technology	New technology
Variety	Pardo Pastusa	ICA-Guantiva
Seeding density*		
Rows/ha	90	100
Holes/100 meters of row	200	250
Tubers per hole	2-3	1
Fertilization		
10-30-10 (at seeding)	750 kg	700 kg
Pest control		
Flea beetle (*Epitrix* spp.)	Poor control	Control
Cutworm (*Agrotis* spp.)	Poor control	Control
Late blight (*Phytophthora infestans*)	Poor control	Control
Production		
Average yield	11,280 kg/ha	17,040 kg/ha
Standard deviation	9,710 kg/ha	9,610 kg/ha
Net earnings	US$313/ha	US$406/ha

*Seeding density was increased (more holes/ha). The tubers per hole were reduced because the new variety had bigger tubers and more buds; as a consequence, only one tuber was required per hole to produce the same amount of branches.

ADOPTION OF TECHNOLOGY

Credit Acceptability

Demand from potato farmers for ICA-Caja Agraria supervised credit was quite substantial. Requests were greater than the amount available, and several producers found they were unable to obtain the desired credit for potato production. However corn producers applied for only a part of the available credit—a surprising result, given the ability of the new technology to more than treble net income.

Adoption Rates

Table 9.3 shows the adoption rates for the different components of the new technologies. For the traditional potato farmer who was already applying more than the recommended amount of fertilizer and controlled pests to a certain extent, the only change in technology was use of improved seed. Change in seeding density was recommended only when the improved variety was used. This means that the overall adoption rate for potato technology was

Table 9.3. Adoption rates of new potato and corn technology

Recommendation	Potato*	Corn
	(%)	(%)
Variety	69	83
Seeding density	50	84
Fertilization	100	55
Pest control	64	83

*Unpublished data provided by G. Escobar P. (1973).

the rate observed for the variety component, i.e., 69 percent—not an excellent value for the first year of the recommendation, but certainly promising.

In the case of corn, while adoption rates were generally satisfactory, some problems existed with fertilizer adoption. Adoption of the variety, seeding density, and pest control components (which required substantial changes in cultural practices compared to the traditional system) were actually higher than in potatoes; however, the fertilizer adoption rate was only 55 percent.

Further analysis of fertilizer adoption (Table 9.4) revealed that all farmers applied the first application of 10-30-10 (although generally at less than the recommended levels), but only 17 percent applied some urea at seeding. A second urea application, 40-50 days after seeding, was attempted by only 48 percent of the farmers, although almost all applied less than the recommended level.

Further studies, carried out with the farmers who had attended the new corn-technology extension field days but did not accept credit, revealed two subgroups whose adoption behavior differed significantly, especially after one year of limited testing of the recommendations on their farms.

One group comprised farmers who basically adopted the recommendations on seeding density, pest control, and hybrid seed, but not fertilizer. The other group accepted seeding density and pest control but continued to use traditional seed varieties and failed to apply fertilizer. Different reactions to

Table 9.4. Fertilizer adoption in corn production

Fertilizer recommendation	None applied	Quantity applied less than recommended	Quantity applied equal to recommended	Quantity applied greater than recommended
	(%)	(%)	(%)	(%)
10-30-10 (at seeding)	0	65	9	26
Urea (at seeding)	83	4	0	13
Urea (at 40-50 days)	52	43	0	5

these recommendations were obtained by the two subgroups. The first group (who used hybrid seed) were dissatisfied with the production, which was similar to that previously attained with local varieties without the other parts of the recommendation. In addition, producers said the corn was of inferior quality—i.e., the taste and ease of preparation of *arepas* (a local bread) and *cuchuco* (a local soup) changed considerably. As a consequence, the group returned to seeding traditional varieties the next year. In contrast, the second subgroup (who used regional varieties) were quite satisfied with the production and continued using the pesticides and seeding density recommended by the project.

Identifying Causes for Partial Adoption of Technology

In general, no major input shortages were encountered for the application of the new corn technology. In the case of the new potato technology some problems existed because the supply of improved seed was insufficient to satisfy the request of all producers. For this reason the project staff considered that the 69 percent adoption rate underestimated the potential if improved seed had been available in sufficient quantities.

To understand more clearly how the small farmer perceived the new technologies and why fertilizer adoption was higher in potatoes than in corn, each was compared with its respective traditional production system, thereby identifying and quantifying criteria that were considered to be important for the small farmers (Tables 9.5 and 9.6).

Changes in Production and Net Gain

Very little is gained through explanation of the higher adoption rates of potato technology by simply comparing the changes in production and net gain generated by the new methods. While the new potato technology increased production by 51 percent and net gain by 30 percent, corn increased these values 202 and 253 percent respectively. If changes in production and net gain are used as criteria to classify technologies, no doubt exists that the new corn method is better than the new potato method. Why, then, did the Cáqueza small farmer fail to recognize this fact? Apparently the saying that "a good technology sells itself" does not apply in all cases.

Changes in Investment Requirements

The new potato technology increased total investment requirements by US$59 (12 percent) and cash costs for material inputs by US$28 (10 percent). The new corn technology increased those values by US$144 (170 percent) and US$131 (575 percent) respectively.

These two comparisons show that the new technology in potatoes, in spite of the fact that it did not generate spectacular increases in production and income, did not require substantial changes in variable costs; it was well

Table 9.5. Comparisons of the traditional technology with the new potato technology

Criteria	Traditional technology	New technology	Percentage change
Production (kg/ha)	11,280	17,040	51
Net gain (US$/ha)	313	406*	30
Total costs (US$/ha)	481	540	12†
Cash costs for material inputs (US$/ha)	285	313	10†
Land returns (US$/ha)	242	377	56
Labor returns (US$/man-day)	2.04	2.30	9
Returns to total investment (US$/US$)	0.65	0.75	16
Returns to cash invested in material inputs (US$/US$)	1.09	1.30	19
Probability that gross income will be less than total cost	0.32	0.21	−34
Probability that gross income will be less than cash costs for material inputs	0.21	0.04	−81
Expected value of the loss function†† using total costs (US$)	218	215	−1
Expected value of the loss function†† using cash costs for material inputs (US$)	133	132	−1

*Net gains for the new technology did not rise proportionately to production increases because the market price for the new varieties was below that of the traditional varieties.

†Unfavorable change for the small farmer.

††For a derivation of the expected value of the loss function, see the Appendix.

within the scope of existing cash limitations. On the other hand, the new corn technology was not only more expensive than that for potatoes (in total and cash values) but also implied major changes in investment requirements with respect to the traditional method.

The traditional corn methods used labor and land production factors predominantly (75 percent of the total investment), whereas the new corn technology required that 62 percent of the total investment be spent on material inputs (fertilizer, new seed, and pesticides), thereby increasing the costs of these factors by an impressive 575 percent.

The new corn technology obviously represented a major change in the traditional production system of low cash investment, a change that perhaps was impractical, as it probably exceeded the limitations of the farmer. To understand clearly what this means for the small corn farmer, it would be useful to compare these requirements with the average disposable income available in the region. According to the data obtained in the Cáqueza Proj-

Table 9.6. Comparisons of the traditional technology with the new corn technology

Criteria	Traditional technology	New technology	Percentage change
Production (kg/ha)	907	2,740	202
Net gain (US$/ha)	58	205	253
Total costs (US$/ha)	85	229	170*
Cash costs for material inputs (US$/ha)	21	142	575*
Land returns (US$/ha)	95	241	155
Labor returns (US$/man-day)	3.07	5.10	73
Returns to total investment (US$/US$)	1.68	1.89	13
Returns to cash invested in material inputs (US$/US$)	3.75	2.44	−58*
Probability that gross income will be less than total cost	0.28	0.13	−53
Probability that gross income will be less than cash costs for material inputs	0.12	0.06	−50
Expected value of the loss function using total costs (US$)	37	78	111*
Expected value of the loss function using cash costs for material inputs (US$)	3.25	53	1530*

*Unfavorable change for the small farmer.

ect (Escobar, 1973) average income per hectare was US$235 (i.e., US$190 from agricultural production and US$45 from off-farm employment).

From another study made in the Cáqueza Project (Shipley and Swanberg, 1974), based on 259 family interviews, it was found that the average value of food consumption (home produced and purchased) was US$222/ha. Subtracting food expenditures from total income, it can be shown that the small farmers have only US$13/ha in cash for investment in the production system. The data from these studies were obtained and processed independently; hence the results are not entirely comparable. Nevertheless, this gives a rough estimate of the amount of cash available for production investment on small farms.

The traditional corn technology required a cash investment of US$21/ha compared with US$142/ha for the new technology. This shows that to adopt the new technology, the small farmers must enter the credit market. And even then, once credit was obtained, it was certain that part was used for other expenditures not necessarily related to corn production, with the result that the quantity of fertilizer (the most costly input) applied was substantially less than the levels recommended.

This could be one of the possible explanations as to why the new potato technology, which did not introduce major changes in investment requirements, was much more easily accepted than the one for corn.

Changes in Returns to Factors

The new corn technology was superior to that for potatoes with respect to returns to land (155 percent increase for corn against 56 percent for potatoes), and the same was true for returns to labor (73 percent increase for corn against 9 percent for potatoes). A small difference existed in favor of the new potato technology for returns to total investments (16 percent against 13 percent) and a much bigger difference, in the same direction, for returns to cash invested in material inputs. While the new potato technology increased the returns to cash invested in material inputs 16 percent, the new corn technology reduced these returns 58 percent.

Given the very limited availability of cash to the small farmer, the fact that the new corn technology reduced returns to this factor could be another explanation for its low rate of adoption.

Changes in Risk Levels

Four different measures of risk were included in the analysis. Two measures were the estimated probabilities that gross income was less than total costs or that it was less than cash costs for material inputs. The other two were the expected values of the loss function from using either total costs or cash costs for material inputs.

The first two measures showed similar results for both the corn and potato technologies. Both substantially reduced the probability that gross income would be less than either total or material input costs. But results were completely different when the expected value of the loss function was used as a measure of risk. The new potato technology had virtually the same expected value of the loss function (for both total costs and cash costs), while the new corn technology increased the expected value of the loss function by 111 percent when total costs were used and by an astonishing 1530 percent when cash costs for material inputs were used. This indicates another possible reason why the corn farmers did not adopt the new technology to the same extent as the potato farmers.

Changes in Labor Requirements

A labor-use study carried out in the area of the Cáqueza Project (Swanberg and Escobar, 1975) demonstrated that labor utilization is not constant throughout the year (see Figure 9.1). In spite of the fact that a high rate of unemployment for the entire region was observed during peak labor-use

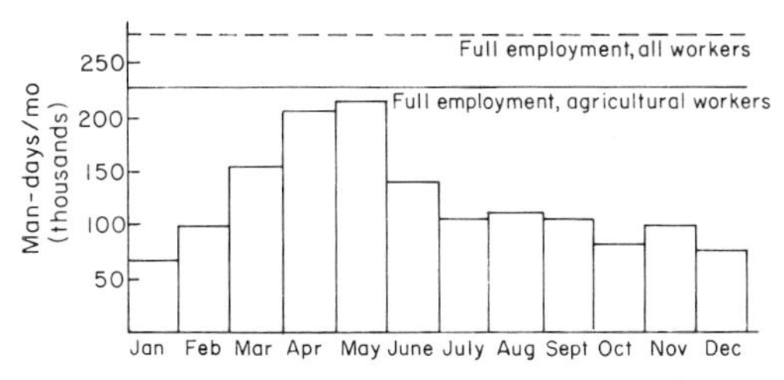

Fig. 9.1. Labor use in the Cáqueza region.

periods, which correspond to the time of corn seeding and first weeding, a point of labor scarcity was reached. This fact led to the discovery of perhaps another justification for the different rates of adoption of the corn recommendation for the first and second fertilization. It was found that the extra labor requirements that would be necessary for the second urea application (see Figure 9.2) exactly coincided with the month of higher labor use in the region. Corresponding labor requirements for harvesting the increased yield did not tax the labor supply schedule because harvesting was spread out over several months during a period of relative abundance of labor.

In Figure 9.3 it can be observed that it would be impossible for all corn farmers in the area to adopt the new technology with two fertilizer applications, since the total labor requirements would substantially exceed the existing supply.

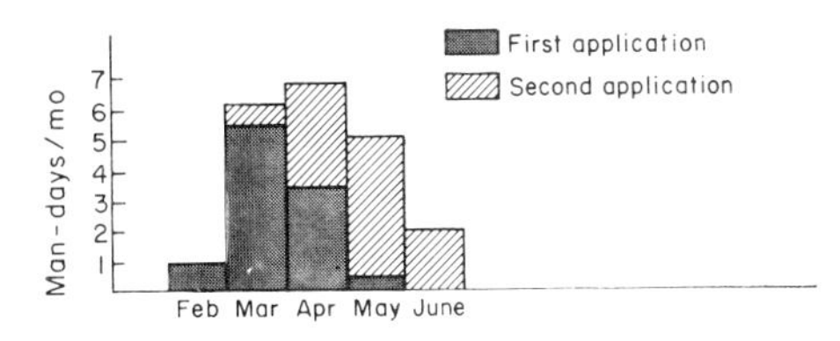

Fig. 9.2. Labor requirements for fertilization in the new corn technology.

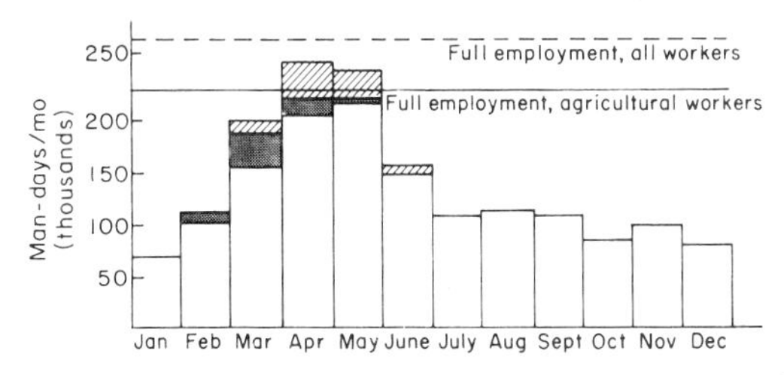

Fig. 9.3. Labor use in the Cáqueza region if all farmers fertilize their corn.

In contrast, the impact on labor supply and demand generated by the new potato technology was inconsequential. The only increase in labor requirements for potatoes was in harvesting, and this was spread through the year, with the lowest use in April and May.

We have presented four possible explanations for the lack of enthusiasm for the new corn technology, while at the same time justifying why the new potato technology did not experience rejection. First, the new corn technology required a significant increase in total costs and, more important, cash costs. Since cash costs exceeded cash availability levels for these farmers, they would be forced into the local credit market if they were to adopt the new technology. Second, the increase in cash costs was produced without generating increased return. Third, the expected value of the loss function (a risk measure) increased tremendously with the new corn technology. Fourth, the labor requirement for the second fertilization in corn was concentrated in the only two months where a possibility of full employment existed, i.e., when surplus labor was unavailable.

The situation with respect to potatoes was altogether different. Cash cost increases were modest and risk levels did not change, which means the new technology did not force a change in the existing system for providing cash. Nonetheless, if it was required, the potato farmer already had a line of credit from either institutional or informal sources; hence if it was necessary to increase borrowings, it would not be traumatic. The extra labor required came during the period when labor was available, which means the labor requirements presented no problem.

Finally, it must be recognized that the potato producers, from their experiences acquired over the years, already knew it was necessary to apply substantial quantities of fertilizers to obtain respectable production levels and incomes. For these reasons when they applied for credit, they used the money almost entirely for the purchase of fertilizers, which they applied in quantities almost equal to those recommended by the new technology. Corn producers were just the opposite: they had never applied fertilizer to their corn crops. This lack of experience in fertilizer use, coupled with low levels of available cash, undoubtedly helps to explain why the credit some producers received was not used entirely for the purchase of material inputs.

Corn and Potato Production Functions

Farmers who used the hybrid corn seed without fertilizer expressed dissatisfaction with the results they obtained compared to those when traditional varieties were used. Hence, the production functions for the traditional system and for the new technology were analyzed.

The equation was as follows:

$$Y = 1.90 - 0.76V + 1.328N + 0.52NV - 0.228N^2$$

where Y = production of corn in tons per hectare
V = traditional (0) or hybrid (1) seed
N = nitrogen fertilizer in units of 100 kg/ha

This equation was obtained in 1975 by pooling data from several studies. For this reason, the recommendations that could be derived from it do not necessarily coincide with the ones the project previously made.

The corn production functions are presented in Figure 9.4. It can be observed that production of hybrid seed is not superior to production of traditional varieties over the entire range. To the left of the crossing point the traditional varieties outperform the hybrids, while to the right of this point the production from hybrid seed is greater. Stated in another way, the hybrid seed performed better than the traditional varieties only when the complete technological package was applied. If the complete package was not used with the hybrid seed, the small farmer could be relatively worse off than if he remained with the traditional variety. A strong positive interaction term between fertilizer and seed variety was found in the new technology production function.

Similar types of production functions were estimated for potatoes. It was found that at zero level of fertilization with nitrogen and phosphorus the traditional Pardo Pastusa variety outperformed the improved ICA-Guantiva. But since no potato farmers in the region failed to apply substantial amounts of fertilizer, extrapolations to zero levels are somewhat dubious. Hence the phenomenon encountered in corn production, where improved seed was inferior to the regional variety at low fertilization levels, adds yet another explanation for why farmers who partially used the recommendation, with little or no fertilizer, preferred the traditional varieties to the hybrids.

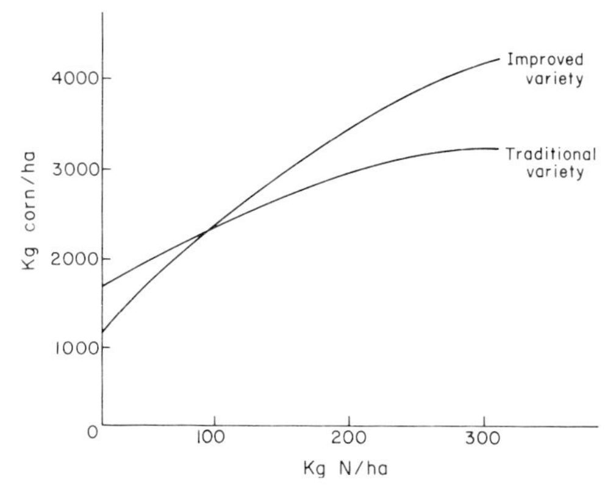

Fig. 9.4. Corn production functions.

ADJUSTING REQUIREMENTS TO LIMITATIONS

From the results of the studies described here, project personnel realized that the requirements of the new corn technology exceeded the limitations facing the small farmer. If the technology was only partially adopted, it would not produce substantial benefits.

In attempting to find a solution, project researchers analyzed various possibilities and discussed these ideas with the producers and the rest of the staff. Risk was found to be one of the most important limitations to farmer adoption of the new corn technology. We have preferred not to make the distinction between risk and uncertainty. Hence the word risk has been used in its generic form.

Three types of risk were identified: production, marketing, and institutional. They are defined as follows: (1) production risk involves the variations in the quantity produced dependent upon causes outside the control of the farmers, such as differences in soil quality, climatic variation during the growing period, etc.; (2) marketing risk includes the presence or lack of a market, price variation during the harvest period and from year to year, changes in demand, etc.; and (3) institutional risk is the presence or absence of seed, fertilizer, and other material inputs at the proper time and the timely availability of credit, transportation, technical assistance, etc.

It was also conceptualized that

1. Each crop and/or subsystem of production requires small farmers to assume risk.
2. A high positive correlation exists between risk, investment levels, and average expected income levels for any agricultural production activity.
3. The small farmer has a certain capacity to absorb risk. This capacity is determined by present income level and/or wealth and not by expectation of increased income.
4. The small farmer is efficient, in the sense that the entire capacity for risk assumption is used and distributed among the cropping enterprises chosen for each growing period.

Under this conceptualization the new corn technology (in spite of its possibilites for substantially increasing expected production and net income) was beyond the adoption possibilities of the small farmer. It required a substantial increase in cash outlays and was too risky.

The Corn Production Plan

Based on this analysis of the cash-risk limitations of the small farmer and the existing labor limitations, the project staff decided to design a corn production plan that modified some of the restrictions previously limiting adoption (Zandstra and Villamizar, 1974).

CHANGING THE CASH AND RISK REQUIREMENTS. The new plan first guaranteed that if production was below a certain level per hectare, the small farmer would not have to pay for the material inputs received for the new technology, thereby reducing production risk. Second, in cases where production exceeded the level fixed as minimum yield (where repayment for credit was not required) the extra quantity produced would be divided in equal parts between the producer and the plan organizers. Since these payments would be made in specie, some market risks for the farmer would be reduced. Third, credit would be given in specie at the time of plan initiation; hence, farmer institutional risks with respect to timing of inputs would be reduced.

The Cáqueza plan was organized through the local producer cooperative, which was originally promoted by the project staff, and operated in the following way:

1. Each participating producer paid a subscription fee equivalent to US$10/ha and invested $US102 in land and labor.
2. The cooperative gave the producers all the seed, fertilizer, and pesticide required for the crop according to project recommendations (US$106). These inputs were given in proportion to the area the farmer was going to seed.
3. At the time of harvest any production equal to or less than 800 kg/ha remained with the producer; up to this point he did not have to return the value of the material inputs to the cooperative.
4. If production was greater than 800 kg/ha, the excess was divided in equal parts between the farmer and the cooperative.
5. The cash equivalent net gain to the producer was $US168, to the cooperative $US48.

CHANGING LABOR REQUIREMENTS. Since it was found that labor was scarce during the seeding and weeding periods for corn, an attempt was made to modify labor requirements for fertilization without reducing potential yield increments. In nine experiments carried out in 1973 and 1974, it was found that one fertilization application 25 days after seeding produced yields similar to those when two applications were used.

In addition, tests were carried out to measure the yield impact due to increased weedings at intervals of 20 days. Yields increased 300 kg/ha when two additional weedings were added to the two already practiced over the crop production cycle. As a consequence, the recommendation now includes the two extra weedings.

In spite of the changes made, some supply problems still remained in regard to fertilization labor, though much less so than when two applications were made. The extra weedings had no impact on full employment because they occurred during June and July, when a fair amount of slack was present.

Acceptability and Results Obtained with the Production Plan

The Cáqueza plan was put into practice during 1974 and was very well received by the producers, in spite of the fact that it was introduced during the major seeding period and that as a result only a few producers had good land still unplanted. During that year, 23 producers participated in the plan. The comparison of earnings and risk calculations between use of the traditional technology and the new modified technology with the proposed production plan is presented in Table 9.7. The requirements and limitations were modified substantially, thereby increasing the possibility that the small farmer would adopt the new technology. Although net income to the farmers under the plan was less than that obtained from the new technology without the plan, the plan nevertheless was accepted without any resistance from the producers and was preferred by them to the institutional credit that offered

Table 9.7. Comparison of the traditional technology with the new technology in corn production, modified by the Cáqueza plan

Production information	Traditional technology	New modified technology*	Percentage change
Production (kg/ha)	907	1770	95
Net gain (US$/ha)	58	168	169
Total costs (US$/ha)	85	112	32†
Cash costs in material inputs (US$/ha)	21	31	48†
Land returns (US$/ha)	95	204	115
Land returns (US$/man-day)	3.07	4.36	42
Returns to total investment (US$/US$)	1.68	2.50	49
Returns to cash invested in material inputs (US$/US$)	3.75	6.42	71
Probability that gross income will be less than total costs	0.28	0.04	−85
Probability that gross income will be less than cash costs for material inputs	0.12	0.01	−92
Expected value of the loss function using total costs (US$)	37	39	5†
Expected value of the loss function using cash costs for material inputs (US$)	3.25	0.50	−85

*All values correspond to those received by the small farmers participating in the plan.
†Unfavorable change for the small farmer.

greater net profits with the adoption of new technology but did not share any of the risks.

The new technology with the plan generated only a 95 percent yield increase and a 169 percent net gain increase compared to figures over 200 percent for both these variables for the new technology without the plan. However, farmer costs increased very little, which was one of the objectives. Land and labor returns also declined somewhat, but returns to total farmer investments almost quadrupled, and returns to cash invested in material inputs improved substantially because instead of being reduced (−58 percent) with the new technology without the plan it was increased (71 percent) with introduction of the plan. However, the key factor in analysis was that the expected value of the loss function with respect to cash costs for material inputs was reduced 85 percent by employing the plan, compared to an increase of 1530 percent for the new technology without it.

Results obtained after the first year of experimentation showed that average harvests were inferior to those expected. However, the producers doubled their incomes and the cooperative had only a small loss. In addition, it was found that the administrative structure of the plan was difficult to manage because it forced project personnel to visit each crop at the harvest time to observe yield levels directly (1974 was a particularly bad production year for corn, with an estimated local traditional production level of only 600 kg/ha instead of 907 kg/ha as observed in other years.)

During 1975 the plan was repeated, but with several modifications to improve management. The principal change was that instead of dividing half the production in excess of 800 kg/ha between the cooperative and the farmer, the producers would have the right to all production equal to or less than 800 kg/ha, the next 900 kg/ha would be given to the cooperative to pay for the material inputs, and anything in excess of this amount would be kept by the producers. In cases where the producers estimated that the production was to be less than 1700 kg/ha (800 kg/ha for the producers and 900 kg/ha for the cooperative), the producer could solicit an inspection to prove that the harvest was really less than that amount. If this was so, the producer would only have to pay the quantity that was in excess of 800 kg/ha. In case it was observed after an inspection that production was in excess of 1700 kg/ha, the excess would be divided equally between the cooperative and the producers. With this modification to the plan it was estimated that inspection time and costs would be reduced by 80 percent during 1975.

In spite of these changes 39 farmers participated in the plan in 1975, thereby exhausting all the available funds for this program; in the process they completely adopted the new technological package. Achievement of a primary objective of the plan, adoption of the new technology, has been demonstrated by analyzing the adoption rates of the participating farmers—greater than 90 percent on all components in 1974 and approximately 95 percent in 1975.

In addition during 1975, with very similar results, a similar investment plan was developed for onion production with the participation of 14 farmers.

ROLE OF ECONOMIC ANALYSIS IN DESIGN OF NEW TECHNOLOGIES

Socioeconomic and Biophysical Requirements and Limitations

From the studies referred to earlier, the project personnel were able to visualize the requirements of new technologies and limitations of the small farmers. While the biophysical requirements and limitations were very easily understood by the agronomist (and as a consequence led to the early formulation of activities designed to adjust the new technologies to the biophysical local conditions), acknowledgment of the existence of socioeconomic requirements of the new technology and socioeconomic limitations of the small farmer took much more time. Nonetheless, the most common socioeconomic requirements and limitations are now being recognized.

New technology almost always requires an increase in the use of resources. The basic resource is capital, as reflected in material inputs and labor. Land for the small farmer in Colombia is considered relatively fixed. Hence, the new technologies are either laborsaving (mechanization) or land augmenting (fertilizer, new seeds, pesticides, or herbicides), both of which require increased capital expenditures. For such cost increments to generate higher incomes, an efficient marketing system must be in operation and adequate demand must exist. It is also recognized that as capital usage rises, various types of risk correspondingly increase.

The limitations of farmers, then, are determined by cash reserves, capital access, and the level of the opportunity cost of capital vis-à-vis capital return rates; the ready availability of the necessary material inputs and their local prices in comparison to the yield increases they generate; the amount of labor available for each activity in the production process, along with the degree to which real labor costs and returns are represented by the prevailing wage rate; the presence of elastic or inelastic derived product demand functions and the degree of efficiency in which the marketing system functions; and farmer capacity to absorb risk as determined by wealth and income levels.

Types of New Technologies

From this analysis it is possible to identify four types of new technologies. The requirements for technologies of type I fall within the socioeconomic as well as biophysical limitations of the small farmer. If this technology increases farmer income, it will be very easily adopted. An example of this type is the new potato technology, which only requires a change in seed variety to increase yield and income over 30 percent.

Type II technologies are physically possible but are beyond the socioeconomic limitations of the target farmer groups. An example of type II was the new corn technology, which physically increased yield threefold but also increased risk beyond the absorptive capacity of the Cáqueza farmers.

Type III technologies are socioeconomically possible, but their biophysical requirements surpass the limitations of the region, e.g., other potato varieties that performed very badly in the Cáqueza region.

Finally, type IV technologies are those whose set of socioeconomic and biophysical requirements are greater than the set of small-farmer limitations. An excellent example of this type is some of the corn hybrids tried in the Cáqueza area that required substantial changes in cash outlays for fertilizer but were unable to produce because they did not adapt to regional climatic conditions.

Agricultural experiment stations are able to generate any of these four types. By doing agronomic research at the farm level, the rural development projects are able to determine if the biophysical requirements of the new technology are within or beyond the limitations of the region. By concurrent socioeconomic evaluation, the farmers' sets of limitations can also be derived. In this way it becomes possible to identify the emerging technologies by type (Fig. 9.5). If the new technology is either type III or IV, it should not be recommended because it does not adapt to the region and hence would be counterproductive to the goal of increasing rural welfare. In cases where the new technology is type I or II, biophysical research and adjustment alone is insufficient to determine specifically which type it is.

		Socioeconomic	
		Requirements within limitations	Requirements beyond limitations
Biophysical	Requirements within limitations	I (Potatoes)	II (Corn)
	Requirements beyond limitations	III	IV

Fig. 9.5. Types of technology according to biophysical and socioeconomic requirements and limitations.

A socioeconomic understanding of the demands made by the requirements of the new technology being generated, juxtaposed against the socioeconomic limitations that face the small farmer, is just as necessary as the basic knowledge of the biophysical requirements under local conditions. By identifying these socioeconomic requirements and limitations, it becomes possible to identify new technologies as type I or II. In the first case, the technology can readily be extended to small farmers and easily adopted by them (the case of potatoes in Cáqueza). In the second case, it may be necessary to design production plans (as for corn in Cáqueza) that modify the socioeconomic requirements and limitations to improve adoption rates. In contrast, if type II new technologies are communicated to farmers without the introduction of special production plans, only those farmers with lower socioeconomic limitations (higher wealth and income levels) would adopt. This repeats once again the all too well-known situation documented in the critical literature of the green revolution in many parts of the world, wherein the new technology is adopted only by the larger land-owning farmers.

Interrelationship between Socioeconomic and Biophysical Requirements and Limitations

CÁQUEZA PROJECT LEVEL. In conceptual and diagrammatic form, Figure 9.6 demonstrates, the role economic analysis has played in the Cáqueza Project. The diagram shows that it was necessary to identify and understand how structural limitations reduce the small farmer's capacity to adopt new technology. In the case of corn (which was found to be beyond the farmer's present capacity for adaptation) a program such as the corn production plan was needed to modify the socioeconomic requirements and, in such a way, place it within the socioeconomic limitations. We consider that these requirements and limitations, both socioeconomic and biophysical, must be studied and observed at the farm level in an integrated form (before any potential new technology can be extended to the small farmers) if any substantial adoption rate is to be expected.

Figure 9.6 shows that the starting point should be the present conditions of small farmers. At this point it is recognized that a set of specific socioeconomic limitations (L_{se}) and a set of specific biophysical limitations (L_{be}) exist. The first task is to experiment with the new technology under local conditions. This technology has a set of specific socioeconomic requirements (R_{se}) and a set of specific biophysical requirements (R_{be}). For adoption or change to take place, it must be shown that R_{se} and R_{be} fall within L_{se} and L_{be} respectively. Only if these conditions are met should this technology be released to the farmers. If this is not the case, the requirements and/or limita-

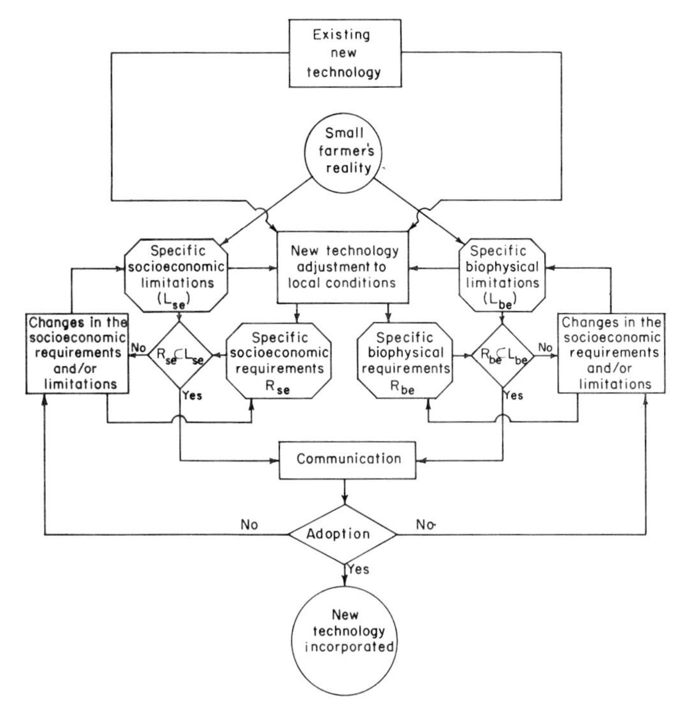

Fig. 9.6. Socioeconomic and biophysical analysis in the adoption of new technology.

tions must be modified (to ensure that both sets of requirements are included in both sets of limitations) before the technology is communicated to the farmers of the region.

EXPERIMENT STATION LEVEL. Extending the same principle to the institutions in charge of generating new technology (see Fig. 9.7), we believe that a careful scrutiny of small-farmer reality must be carried out directly or in collaboration with the existing rural development projects in the area of influence. The identified specific biophysical and socioeconomic limitations of the different regions should be combined to generate *general* limitations, both biophysical (L_{bg}) and socioeconomic (L_{sg}), and this information should be used in guiding the generation of new technology. Once available, the general requirements, biophysical (R_{bg}) as well as socioeconomic (R_{sg}), must be checked with the L_{bg} and L_{sg} of the small farmers. If the requirements surpass these limitations, more research should be carried out to identify new technologies that have requirements within the limitations of the farmers. On the other hand, if the requirements of the newly generated technologies are included within the general limitations, these technologies should be made

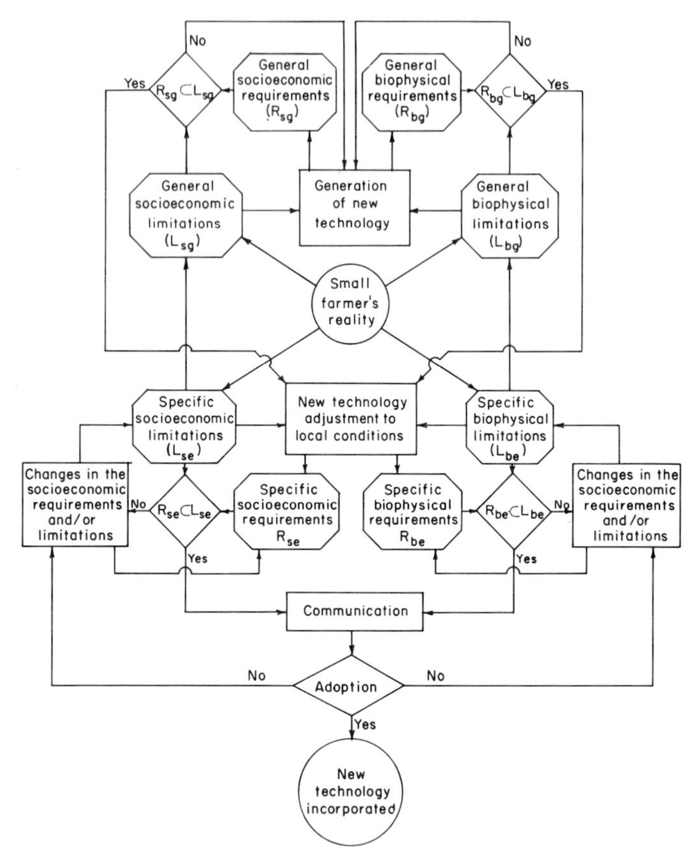

Fig. 9.7. Socioeconomic and biophysical analysis in the generation of new technology.

available to the rural development projects to carry out the necessary adjustments to the specific characteristics of the region. This schematic design has been developed to show that the whole process must start by considering the realities of the small farmer.

The authors feel that this is the complete process required to generate new technology for the small farmer, where economic analysis plays as important a role as any other. If this process is not followed, the degree of farmer adoption will be minimal. As a consequence, the history of failures in generating adequate technologies for small farmers will be repeated; as in the past, the major beneficiaries will be only those farmers with the most resources.

CONCLUSIONS

The role corresponding to economic analysis, from the generation of new technology to adoption by small farmers, is that of identifying the socioeconomic limitations that may restrict adoption and performing a socioeconomic analysis of the requirements of the new technology to determine if they fall within the existing set of limitations presently facing farmers. In cases where requirements are greater than existing limitations, economic analysis must specify how the requirements or the limitations can be modified, so that the new technology can be adopted.

Unfortunately, knowledge in this field is limited. We are clearly unable to specify all the studies that are required ex ante. We consider that this is a continuous process in which many mistakes will be made. In the case of Cáqueza, ex post analysis in the second year became ex ante analysis for the third and permitted the project staff to correct some early errors. One thing is certain: the accumulation of this type of experience will assist in determining what kind of research is required to generate new technologies for small farmers.

The key point is that professionals from biophysical and socioeconomic fields must collaborate in identifying reality by defining farmer limitations and requirements of new technologies. We believe that this interaction through joint research by multidisciplinary teams at research stations and in field projects is one of the best ways, perhaps the only way, to achieve this goal.

APPENDIX

Mathematical Derivation of the Expected Value of the Loss Function

There are three reasons why the expected value of the loss function is used as an estimate of risk for the small farmer. Two are based on the importance that the negative extreme of the net gain function (to the left of the equilibrium point where net gain equals zero) has for the farmer.

1. The estimate takes into account the value of production costs in the sense that this measurement can compare technologies with equal average gains and variances but different costs.
2. In addition, it permits classification of technologies whose accumulated density functions intercept below the point of equilibrium.
3. Finally, the expected value of the loss function can be expressed in monetary terms, which facilitates comparison of production alternatives in different regions. The possibility of adding the expected loss values also

permits comparison between combinations of alternatives and use of this measurement of risk in the development of linear programming analyses.

Let $f_q(y)$, the probability density function of yields y for a crop in a given zone f_q, be defined as

$$f_q(y) \sim N(\mu, \sigma^2)$$

where f_q is a normal distribution with a mean μ and a variance σ^2.

Let the loss function f_l be defined in terms of costs, yields, and prices:

$$f_l(y) = C - yp$$

where C = production costs (total or cash for material inputs)
y = yield in kilograms
p = the price per kilogram of product

In this study the specified loss function is not an actuarial function in the sense that it does not accept negative values (see Halter and Dean, 1971).

The loss function is then defined as

$$f_l = C - yp \quad \text{for} \quad C > yp \quad \text{or} \quad y < C/p$$

$$f_l = 0 \quad \text{for} \quad C \leq yp \quad \text{or} \quad y \geq C/p$$

Hence, the expected value of the loss (l) will be

$$E(l) = \int_{-\infty}^{+\infty} f_l(y) f_q(y)\, d_y$$

or

$$E(l) = E(C - yp) \quad \text{for} \quad y \leq C/p$$

or

$$E(l) = C - [p\, E(y)] \quad \text{for} \quad y \leq C/p$$

or

$$E(l) = C - \{p\, E[y \mid y \leq (C/p)]\}$$

The above calculation assumes that product price and production costs are independent of yield.

In this way the expected value of the loss can be calculated from production costs, product prices, and the expected value of the truncated normal distribution for yield.

COMMENT / *Julio A. Penna*

This chapter is a good example of the role that institutions can play in influencing the adoption of new agricultural technology. The Cáqueza Project allowed the risk associated with production and prices to be shared between an institution and the farmers through assistance in buying the excesses (in the case of overproduction) or exemption from the payment for inputs (in the case of low production). At the same time, this allowed the participating farmers to foresee a more favorable economic outcome.

The budget analysis used by the authors has the advantage of being easily applied to compare the economic consequences of different technological alternatives. However, the method is less advantageous when various technological packages are compared simultaneously.

The authors conclude that the low use of fertilizers for corn was due to high costs, and that "part of the credit was used for other expenditures not necessarily related to corn production." This last conclusion has not, in my opinion, been appropriately defended, since it is probable (although not stated) that the credit was directed to seed, additional labor for higher seeding density, or control of insects. Given the relative prices of inputs in the technological package, it is probable (subject to empirical evidence) that the practices have a comparative advantage with respect to fertilizer use. Note that in the case of corn, the percentage of adoption of hybrid seed (83 percent), density of seeding (84 percent), and control of insects (83 percent) is greater than the adoption of fertilizer (55 percent). If this were the case, the farmers would be acting rationally to discriminate against fertilizer, and the net return they could obtain with this incomplete package would be higher than that obtained with traditional techniques. In effect, the authors indicate that some farmers who used the incomplete package were satisfied with the results obtained, although they did not show the same net average profit nor the expected value of the loss function. I believe that a comparison of the economic results obtained using the modified new technology with the results of the incomplete package would permit an evaluation of the impact of fertilizer use within the recommended package.

Based on economic results for the "existing technology" and "new technology" the authors implicitly conclude that the new technology is rejected because the expected value of the corresponding loss function is higher than that for the existing technology. I believe this conclusion could be questioned, considering the trade-off that exists between comparing the probability that the gross expected profits are greater than the total cost for both technologies with the expected value of the loss function. It should be added that the net expected profit is higher in the case of the new technology. As a result I do not see any reason to conclude that the existing technology is preferred to the new one.

In the proposed loss function the total costs of production are taken as fixed. I believe this could introduce a significant bias in this function due to the obvious fact that the costs depend on the level of output.

JOHN L. DILLON

10

Broad Structural Review of the Small-Farmer Technology Problem

But above all, in order to liberate the peasantries, it is necessary to generate intentionality to do so.
CARLOS BENITO (1975)

Small farmers, poor farmers, peasants, subsistence farmers, rural marginals, *minifundistas,* traditional farmers—call them what you will—constitute a real and complicated human problem. In this chapter an attempt will be made to sketch this problem in terms of its possible causes, structure, and dimensions in relation to the role of economic analysis in the design of new technology for small farmers. To date, though many have suggested that particular emphasis should be given to improved technology specifically oriented to small farmers, its design has not received concentrated attention by economists—nor has the question of whether small farmers would be better off with new technology been adequately resolved.

I propose first to briefly discuss the size and some aspects of the nature of the small-farmer problem. Next I will attempt to broadly summarize the more important theories that have been postulated and bear on the genesis and resolution of the small-farmer problem, in turn assessing the implications of these theories for the development of technology. Third, assuming new technology is to be developed for small farmers, I consider in broad terms the question of what ex ante criteria might be used to guide scientists' choices of effort. And finally, I will briefly consider the question of criteria for ex post evaluation of new technology.

This paper was written while the author was on sabbatical leave from the University of New England, Armidale, N.S.W., Australia, in Departamento de Economía Rural, Universidade Federal do Ceará, Fortaleza, Brazil. I am grateful to Alain de Janvry and John Sanders for crtitical comment.

SMALL FARMERS—A PROBLEM OR NOT?

The definition of small farmers, poor peasants, subsistence farmers, etc., has been the subject of endless, if rather pointless, debate and discussion, e.g., Miracle (1968), Powell (1972), Wharton (1969a). For present purposes, the major definitional elements are a local context of population pressure, the practice of cultivation or livestock keeping, a chronic low standard of living involving either absolute poverty or verging on it, a lack of dynamism and endogenous grounds for hope of a better future in the socioeconomic milieu, and reliance to some degree on subsistence production. Others, I am sure, would wish to rephrase this or emphasize some other aspects. But as Wharton (1969a) notes, no matter what exact criteria we might use, we have much the same basic population in mind.

Just how many small farmers there are in the world appears not to be known with any exactitude. Wharton (1969b) suggests that about half the world population is dependent on subsistence agriculture, about 40 percent of total cultivated land is worked by small farmers, 60 percent of all farmers are small, and they account for less than 40 percent of all agricultural output. Rough as these estimates undoubtedly are, they indicate an immense if not overwhelming problem, which is compounded by both external and internal factors. Leaving the external factors till later, I wish to emphasize here the internal factors of the small-farmer subculture and resource base.

No matter where they are found around the world and through no fault of their own, small farmers appear to constitute a subculture with some or all of such negative characteristics (Rogers, 1969; Doob, 1969) as mutual distrust in interpersonal relationships, lack of innovativeness, fatalism, low aspirational levels, lack of deferred gratification, and lack of empathy. Though doubtless reflecting a situation of cultural equilibrium with the socioeconomic environment in which they eke out an existence, these are hardly likable characteristics—nor are they indicative of a willing receptivity for help. At the worst, Why should anyone consider helping such a standoffish lot? And at best, the giving of help will not be easy.

The problem is further exacerbated by the restricted resource base on which small farmers must operate. In general, they have control over only a small area of land that is usually naturally poor or depleted; they have an extremely low level of human capital in terms of education and health with which to work; and they lack the socioeconomic power with which to gain access to "public" and other services or perquisites available to more powerful, better endowed members of their national "society." The quotation marks are necessary because in such circumstances one cannot properly speak of public services or a national society. With such a small resource base to start from, the degrees of freedom available in attempting to solve the problem are limited. Without boots there are no bootstraps by which to pull oneself up!

And not only are there limited opportunities for leverage from outside, but the penalty costs of a mistake for people at the edge of existence are disproportionately severe. Conversely, though marginal enhancements will typically be small in absolute terms, they will be large in relative terms. For example, a US$100 per capita increase in annual net benefits will often correspond to a doubling of income.

So far I have been referring to the small-farmer problem without any attempt to define it. A definition will be biased or oriented by one's view of the causes of the problem. Leaving a discussion of causes till later, it appears the problem is that we have a mass of small farmers throughout the world who are chronically disadvantaged in their standard of living and in their expectations for their children. I see it as a problem of welfare and social justice. Such a view leads to some rather different thoughts (unexplored here) on approaches to resolution of the problem than does the more orthodox economics view. Under this more usual view, as typified by Owen (1966), the problem is one of marshalling small-farmer resources in such a way that they contribute to and participate in national economic growth.

I do not believe the goal is to assist economic growth by making the small farmer modernized or commercial. Indeed, the term "modernize" has probably done more harm than good with its placatory suggestion to the world at large that a quantum jump in small-farmer technology is just around the corner. The primary goal relative to the small farmer is simply to make him better off so that he goes at least some way toward a better realization of his human potential and rights to self determination. Often, lack of national wealth will prevent treatment of the problem as one purely of welfare. Attempted resolution can then only be through trying to ensure that small farmers contribute to and participate in national economic growth.

As a welfare problem I certainly do not see the small-farmer problem as one to which the Pareto principle is in any way relevant. Within the small farmer's personal socioeconomic sphere of decision making, the usual marginal principles of utility maximization would of course apply. The principle of having change only to the extent that nobody is made worse off (or thinks he is) is no more than a fancy argument for the status quo. With it we would never have had the abolition of slavery nor (where they have been introduced) universal education, land reform, or other significant social change.

Should attempts be made to solve the small-farmer problem? I think this is quite a valid question. In purely cost/benefit terms, one might say yes because of a judgment that over the longer term the "have-nots" will overthrow the "haves." Or the answer might be no, because cost/benefit analysis indicates resources could be better spent elsewhere in the economy to assist both the urban and rural poor or just the urban poor alone. Or one might argue that self-preservation (the "lifeboat" theory) dictates not wasting resources on an impossible situation. More esoterically, doing nothing might be argued on the basis that interpersonal comparisons of utility are impossible (supported by the intuitive knowledge that we can get a lot more enjoyment out of an extra dollar than some poor uncultured peasant!); but I believe the question transcends economic accounting. It is simply right on moral or ethical grounds that we should attempt to solve the problem. However, while this is fine as a principle, in fact resources are limited. In most countries it will be impossible to avoid questions of allocation between alternative posibilities, e.g., between programs directed to small farmers or to the urban poor.

Two other general points need to be made about the small-farmer problem. First, there is no feasibility of its being resolved on any significant scale within less than a matter of decades. The resources needed are too great, and the endogenous and exogenous forces favoring perpetuation of the problem are too strong. Second, both between and within nations, the problem exhibits great variability in its agricultural and cultural settings. Particular approaches, such as a specific policy or technology, that move toward a solution in one region are likely to have little relevance to others. What might be right for a semiarid region of Bangladesh is unlikely to be appropriate for a bog in North Ireland. And when we come to a particular region, we will assuredly need to more finely classify the population of small farmers in one way or another.

EXPLANATIONS OF THE SMALL-FARMER PROBLEM

Many theories of economic growth and development have been postulated and bear to varying degree on the small-farmer problem. For present purposes I will concentrate on what I believe are the three more important types of theories, classifying them under the headings of dual-economy models (e.g., Jorgenson, 1969), Schultz's "poor but efficient" model of traditional agriculture (Schultz, 1964; Mellor, 1967), and the theory of unequal exchange or exploitation between the "center" and the "periphery" of the world economy (e.g., de Janvry, 1975; Stavenhagen, 1969; Szentes, 1971).

In outlining these theories, I have neither the space nor the expertise to do them justice. In consequence my treatment is extremely broad. Because he specifically assumes underpopulation, I have omitted Chayanov's theory of peasant economy (Thorner et al., 1966) in which the subjective valuation of family labor effort (not "financial" return) is the key element and against which the worthwhileness of new technology is judged. Chayanov saw resolution of the (Russian) peasant problem via social reorganization based on larger scale cooperative farming.

Broadly, the dual-economy model of underdevelopment assumes the coexistence of two more or less autonomous sectors within a given national economy—one modern, the other backward. The modern sector centers on industry, urban services, and/or export production from large agricultural units. The backward sector consists of small-farmer agriculture with a large degree of subsistence, a low level of technology, and archaic social organization. Within this broad characterization, dozens of dual-economy models have been specified with alternative assumptions about labor supply, wage rates, savings, technology, etc.

The best known such framework is perhaps that of Jorgenson (1969), which encompasses what have come to be known as the classical approach (a fixed real wage rate and a surplus of agricultural labor) and the neoclassical approach (a variable real wage rate and no labor surplus). None of these models explain the existence of the backward or small-farm sector except obversely in terms of saying that with national economic growth the sector will eventually disappear. They take for granted that it exists to begin with. Under certain conditions, however, the neoclassical model implies a "low-level equilibrium trap" involving perpetuation of the backward sector. Escape from this trap requires either an increased rate of technical change in agriculture, the introduction of capital to agriculture, or a fall in the population growth rate (Jorgenson, 1969). Overall, relative to the small-farmer problem, the dual-economy models do not imply an actively malevolent environment. We might say that it is not that the cards have been stacked against the small farmer, but just that the deal has not turned out too favorably. If the game goes on long enough, things will get better. Given time and economic growth, the small-farmer problem will eventually disappear.

Schultz's model concentrates on the small-farmer problem without linking it to general national economic growth. In his view, small farmers operate in a relatively static technological, economic, and cultural environment to which they have become very well adjusted and within which they operate efficiently as economic men. Given their economic nature, for them to break out of their efficient but poor status they need incentives by way of profitable new technology (backed up by the required input supplies and marketing channels) and, over the longer term, institutional change including education. Like the dual-economy models, which also assume small farmers are efficient (but operate with a constant rate of technological change), Schultz's model does not imply a malevolent socioeconomic environment. To continue our poker analogy, the deck is not stacked but the deal tends to be monotonously similar. Manipulation of the kitty is needed to make the game interesting for the small farmer.

In contrast to the dual-economy and Schultzian theories, the theory of unequal exchange or exploitation between center and periphery implies that small farmers operate under an actively malevolent socioeconomic environment. The cards are stacked against the small farmer. This theory has been best argued for Latin America relative to the capitalist center, but there are certainly signs of such a system within the COMECON grouping too.

Benito (1975) has summarized the theory as follows:

> The unequal development between central and peripheral countries [or regions within a peripheral country] is the consequence of a process of capital accumulation based on conditions of unequal exchange between national formations, between urban-industrial and rural-agricultural areas, and between commercial agricultural sectors and peasantries. Unequal development is expressed not only in a lower per person income but also in a more complex pattern of social differentiation within a peripheral social system. Unequal exchange or transfer of surplus value from one sector to the other is made possible by the heterogeneity of the peripheral social systems [which allows an oligarchy at the periphery to form an unholy alliance with the center to exploit the small farmers in the periphery].

From this viewpoint, therefore, agricultural output stagnation and social marginality (i.e., the small-farmer problem) are necessary consequences in peripheral socioeconomic systems. The existence of small farmers and their continuing impoverishment is seen as crucial to sustaining the transfer of surplus value from the less developed periphery of the world to the developed center. Such a theory goes beyond the traditional economic dimension used in the dual economy and Schultzian models. It invokes questions of power and social conflict and sees the problem as one of political economy rather than economics per se as often (erroneously) perceived. Like the Pareto principle, I see the usual narrow definition of economics as a procedure whereby economists insulate themselves from the world's real problems.

Further, under this theory there appears little hope for a solution of the small-farmer problem. Solution would require significant changes in the political domain and distribution of national resources, not to mention the development of a social conscience recognizing that no person has the right to exploit another, either directly or through the anonymity of commercial or other entities. (I would not call a fairer sharing of wealth exploitation of the rich by the poor.)

Doubtless all the theories sketched here contain some degree of truth, and some are or have been truer than others for small farmers in different countries or places. If I had to choose the one most relevant overall for South America, my bias would be toward a somewhat "soft" version of the center-periphery theory of exploitation with its connotation of the carryover, in one way or another, of colonial birthmarks and mechanisms to today's world. But that is judgment only, as I also believe none of the theories are yet scientifically "proven" and indeed may not be amenable to "proof." I use the term "soft" in the sense that I believe better social justice may eventually prevail under the pressures of nationalism and social dissatisfaction at the periphery. Also, it may be that what is involved is more a question of social blindness and short-sightedness among the wealthy than either an active or a necessarily logical conspiracy against the poor.

What are the implications of these theories relative to the role of new technology in assisting solution of the small-farmer problem? To date, agricultural economists have largely oriented their discussion of the implications of the theories, not to farm technology development, but to questions of agricultural policy, particularly in relation to prices and land control.

Schultz's theory of small farmers being poor but economically motivated and able to reap the initial benefits of new technology gives the major role to the provision of new technology for small farmers. Under this theory, a continuing stream of profitable and feasible new technology solves the small-farmer problem. And if we see and can handle the problem as a welfare one, we should be willing to make new technology profitable through subsidies tailored to small farmers. Under Schultz's theory we should therefore press ahead as fast as feasible with agricultural research and technology development for small farmers.

The provision of new technology is also a crucial element in the dual-economy models. These generally assume a constant rate of technical change in the small-farmer sector. The higher this rate of technical change, the quicker economic development and solution of the small-farmer problem can occur. And for an economy caught in a low-level equilibrium trap, an increase in the rate of introduction of new technology provides a means of escape.

Under the center-periphery theory of exploitation, virtually no hope seems to be given for new technology as a direct means of ameliorating the small-farmer problem. If this theory prevails, all the benefits of technical change will be captured by the exploiters except to the degree that they allow token benefits to be held as a means of mitigating social unrest. Specific mechanisms could be depressed wages from employers, increased rents by landlords, higher interest rates by moneylenders, and lower prices by product traders.

But technological change may still have a role to play—albeit over the longer term—as a mechanism through which small farmers (and others) become more conscious of their plight as marginals and hence less acquiescent to exploitation. Indeed, in this sense of increasing social conscience, new technology may have a key role under the center-periphery theory as a means of catalyzing social change. Whether such a hope is empirically justifiable is something worthy of investigation under appropriate conditions, if they can be found. So even under the theory of center-periphery exploitation, I would argue that the search for new small-farmer technology should be pursued but should be recognized as an indirect step rather than a direct approach to rural development and social justice for small farmers.

GUIDES TO TECHNOLOGY DESIGN

Whether at international, national, or regional centers and regardless of one's particular view of the genesis and solution of the small-farmer problem, the search for new technology will continue. What guidelines or criteria might be offered to assist scientists in their choices, so as to best orient research toward technology for small farmers?

The broadest and most important guideline, I feel, is that there should be an explicit commitment of intent to develop new technology for small farmers. I see the basis for this decision in ethics, not in economics. In terms of research per se, such a commitment involves not choosing the easiest of paths to scientific acclaim. It means, for example, field research in unattractive and difficult locations and working on projects that do not necessarily have the highest potential payoffs in improved productivity. It would usually be far easier for scientists to work on technology for modern commercial farmers.

Another broad criterion, somewhat distasteful but nevertheless not irrelevant since research resources are limited, is the "triage principle" as practiced in wartime field hospitals. Small farmers in particular countries or regions would be classified (like human casualties) into three groups: those who will die no matter what; those who, if promptly treated, should survive; and the walking wounded who can look after themselves. Given that agricultural research has to be fairly location specific in its orientation, I am suggesting that small farmers could be classified on a regional basis under the triage principle. For example, an international center might decide to attempt nothing for small farmers in a particular national region because the country concerned had the resources to do the job itself (i.e., the "walking wounded" case) or because the socioeconomic and political situation was so bad that small farmers could get no benefit (i.e., the "die no matter what" case). Likewise, national centers might apply the triage principle to identify particular regions. In applying this criterion, the basis of judgment, *ceteris paribus,* should be whether the small farmer could be helped, not whether work oriented to him would assist national economic growth.

Given a commitment to small-farmer technology and the choice of particular regions of interest, the question of ex ante research guidelines becomes much more specific and difficult.

First, how do we ensure that research is oriented to small farmers? Some degree of overflow to larger farmers is undoubtedly inevitable. This may not be a bad thing. But it could be very self-defeating; e.g., if technological development for a small-farmer crop such as cassava led to its becoming a large-farmer crop. It seems to me there are three ways of ensuring a small-farmer research focus: by choice of crop or crop mixes and particular cultivation techniques (e.g., interplanting); by choice of a specific regional or ecologic orientation; and by aiming for intermediate technology suited to small rather than large farmers (e.g., implements for animal cultivation rather than tractor implements).

Second, it is necessary to have knowledge of current small-farmer technology and how it relates to the farmer's life-style, culture, community needs, and depletion of resource stock. Ideally, we would wish to understand the farmer's socioeconomic environment so that we can meet the constraints on technology imposed by the community situation. The collection of such information is a difficult task and will often suffer from a lack of expertise and empathy on the part of the collectors, not to mention lack of cooperation by the small farmer. But both CIAT and ICRISAT, for example, appear to be setting good examples in this regard (CIAT, 1974; Jodha and Ryan, 1975).

Third, once information has been collected on the current state of the art and its cultural setting (complemented by the opinions of expert professionals), it can be used as a guide to major needs and feasible possibilities—feasible in the sense of what is researchable and what is applicable by the farmer. This question of feasibility of adoption is all important. Paradoxically, the smaller a recommended change is (or the more the new technology is like the old) the greater its chance of adoption and the less its impact on productivity (except perhaps for the notable case of much better seeds that do not require new complementary inputs). The more new technology diverges from the old, the more likely it is to involve problems arising from cultural and community constraints or pressures, subjective riskiness, and problems of required new input availability. In this sense there is a danger in developing packages of technology. As Ryan and Subrahmanyam (1975) suggest, a series of options rather than a package may be best.

Fourth, based on the above arguments, we should aim for fractionally improved or intermediate (rather than advanced or complicated) technology. This includes new crops as long as their technology is not too different.

Fifth, and in parallel with suggestions three and four above, our research emphasis should be tailored toward technology that matches the farmer's resource and financial and climatic environment; e.g., drought resistance is desirable in semiarid areas, and increased labor requirements in periods of full employment are undesirable. And while we might agree that with population pressure the general bias should be toward landsaving rather than laborsaving technology, this should not be taken for granted in the context of particular situations.

Sixth, once the above guidelines have been used as a screening device to narrow the set of possible technological changes to be researched, ex ante assessment (necessarily on a subjective basis) can be made of the likely probability distribution of net benefits at the whole-farm (not the research station) level for each of these possible candidates. On this basis (with allowance made for presumed farmer risk aversion), for the likelihood of research success and for the size of the population that could benefit, choice of research projects and priorities may be made.

Implicit in the above guidelines is a systems-analysis orientation (see, e.g., Fernández and Franklin, 1973). This may be done quite informally or it might be carried out in more formal ways using systems modelling of varying forms and degrees of sophistication (see Ch. 2). At one extreme a fully computerized procedure might be used. This, however, would run the danger of having too many of the characteristics of a black box for the scientists involved and hence have little favor with them. Sensitivity analysis has an obvious role to play, e.g., in determining break-even levels of expected yields to serve as minimal targets for crop improvement.

Also implicit in the suggested guidelines is a strong degree of directed research. Once the guidelines are applied, they lead to a chosen set of projects and priorities. Though these choices are made by the scientific teams involved, the research choices are not free. This may be something of a disadvantage that has to be borne if the best assistance possible is to be given to small farmers.

In applying the ex ante guidelines, what is the economist's role? I see him neither as a dominant nor a subservient member of the scientific team, but simply as an equal member with his scientific colleagues. Most important, the economist like other team members must recognize the dangers and biases likely to arise because their perceptions and values probably differ from those of the small farmers they are hoping to assist.

GUIDES TO TECHNOLOGY EVALUATION

Once research is under way and results begin to come to hand, evaluation can begin. Full evaluation is not possible until farmer utilization or trials of the new technology provide real-world data. Until then, only relatively soft data will be available. But this should not deter the start of evaluation. Indeed, the early evaluation of research station and field trial data will be very important to extension design activities.

Necessarily, ex post evaluation activities will duplicate much of the ex ante activity. Data on research and farmer results will need to be collected and appraised, leading in turn to further research guidance. In this sense, particularly when an ongoing program of research is under way, ex ante and ex post guideline activities meld together in a continuing cyclical process. Perhaps the greatest distinction between ex ante and ex post activites is that ex ante appraisal must necessarily rely much more on synthesized data and hence is far more subjective.

As far as the techniques of economic appraisal to be used in ex post evaluation are concerned, they will encompass the usual gamut of budgeting, risk programming, etc. The implicit approach should be that of systems analysis applied within a whole-farm context at the farmer level and applied also, to judge broader effects, at the community level. I would stress the need for appraisal in the whole-farm context.

Too often in the past, evaluation has been on a single-crop basis that ignores questions of how a particular activity fits into the whole-farm situation. Only in this way can adequate allowance be made for the interdependencies that exist between activities and resource-use possibilities, for the institutional constraints under which the farmer has to operate, and for farmer risk preferences, when as yet there seems little information or consensus on just what criteria guide small farmers in their risky choices. While evaluation in orthodox economic accounting terms is obviously important, evaluation in broader social terms is also necessary.

SUMMARY

To summarize, I see the small-farmer problem as one of welfare and social justice and, if resources were available, would prefer to treat it as such. Our aim should be to improve the small farmer's lot. Whether new technology tailored to farmer use can do this in a significant way and as a sole activity not linked to broader changes is still an open question. Schultz's poor-but-efficient hypothesis and the dual-economy models suggest it could. But the center-periphery theory of exploitation suggests it could not. Nonetheless, the search for new technology for small farmers should be fostered and guided by criteria that ensure that the research focus is really oriented to the development of technology tailored to the circumstances and needs of small farmers.

COMMENT / *Peter B. R. Hazell*

My comments on this chapter are twofold. First, I object to Dillon's claim that "no matter what exact criteria we might use, we tend to have much the same basic population [of small farmers] in mind." Dillon has in mind a class of farmers whose common characteristics seem to be "mutual distrust in interpersonal relationships," "lack of deferred gratification," "lack of empathy," "standoffish," etc. I would argue that we ought also to give some thought to including "nice" people in our definition of small farmers.

Second, I would argue that all we really need to be concerned about with respect to the size of the small-farm problem is that it is *big*—too big to be treated as a welfare problem. Except in a very limited number of countries, such as Brazil or possibly Mexico, the resource and income base simply does not exist to enable massive welfare transfers to alleviate the small-farm problem. More generally, small farmers are going to have to pay their own way in attaining improved living standards, and the design of small-farm technology must be geared to this end.

Dillon considers only the case of new technology aimed at increasing the marketed surplus, and he is correct to highlight the institutional and economic constraints that restrict the opportunities in this direction. I suggest that an alternative or supplementary approach would be to develop new technologies aimed at increasing the productivity of food crops for farm family consumption. These would have to avoid dependence on modern purchased inputs and hence would probably be limited to small-step technologies, such as modified husbandry practices (e.g., better planting densities and weeding practices).

Possibly some "miracle" technologies might also be found to do the job, such as nitrogen fixation in cereals or the introduction of entirely new food crops. In addition to raising living standards among large numbers of small farmers, this kind of approach could also help alleviate the worst kinds of poverty in some countries. While obviously not the long-run solution to the small-farm problem, this approach is certainly more viable in the short run than Dillon's proposed application of the triage principle.

ALAIN DE JANVRY

11

Nature of Rural Development Programs: Implications for Technology Design

During the 1960s, unprecedented growth in gross national product has been achieved in many less developed countries. Simultaneously, income disparities (and sometimes absolute poverty) have increased sharply, thus bringing to bear forcefully the fact that economic growth does not imply equitable development. With most poverty concentrated in the rural sector, interest in the issue of rural development has been aroused among international agencies and national governments. Funds have been committed and attempts made at defining rural development and identifying the essential components of rural development programs. Yet it is blatantly clear that a consistent body of knowledge regarding rural development still does not exist. Because of this programs generally have been neither designed nor evaluated in terms of the socioeconomic process through which rural poverty has been produced and is being reproduced; and pauperization of rural areas in the Third World remains an increasingly untenable social phenomenon.

Establishing a body of knowledge on rural development and identifying the significance of different instruments for rural development require (1) making explicit the overall *process* by which rural underdevelopment has been created and is being perpetuated in Third World countries; (2) typifying the various *structural* conditions within which this process applies, and thus identifying for each the applicable range of instruments for rural development; and (3) determining the *goals* of rural development programs for each particular instance.

Methodological principles to establish this body of knowledge include the need to recognize (1) the *interrelatedness* of the elements of peasant society with the overall economic and social structure at the national and international levels; (2) the *historical* dimension of the process of transformation of peasant society that permits identification of its laws of motion; and (3) the *particular vision* of the world that peasants have, which is determined by the specific conditions under which they live—this vision establishes the meaning of change for the peasant.

Among the many instruments of rural development, technological change has been singled out as one of primary importance and relatively easy manipulation. Indeed, the nature of technology largely determines the fabric of society but, reciprocally, social relations also condition technological innovations and their diffusion. Questions regarding the design of a small-farmer technology for rural development thus cannot be dissociated from the nature of the development process and the structural conditions that prevail nor from the specific goals expected from rural development programs. We show here that identification of the trio of process, structure, and goals provides important guidelines for the nature of the technologies to be designed.

Agricultural development is contrasted between two sharply different processes: (1) the "western paradigm," which consists of transposing today's less developed countries into the urban-industrial model that induced the nineteenth-century "takeoff" of the center economies, and (2) the "rural way" to economic development, where rural development becomes the primary engine of growth. This dichotomy does not coincide with the contrast between capitalist and socialist processes. Here, the "western paradigm" includes the Soviet model, while the "rural way" includes the Chinese model.

These two processes and the particular structural conditions under which they apply, permit identification of a few major types of rural development programs in terms of goals. Each type requires a particular technological basis.

AGRICULTURAL DEVELOPMENT UNDER THE WESTERN PARADIGM

The Western Paradigm

The theory of economic development being followed by nearly all capitalist less developed countries is derived from the history of today's more developed countries. Under this theory, development is equated with urban-industrial growth. The role of agriculture is to generate surpluses for investment in the modern urban-industrial sector and to create a market for its products. Food surpluses permit the release of labor for employment in the modern sector; financial surpluses are extracted by taxation, the terms of trade tilting against agriculture, rent paid to absentee landlords, and voluntary investment of agricultural savings in the modern sector. The demand by the agricultural sector for inputs and consumption goods produced in the nonagricultural sector augments aggregate demand thereby sustaining continued accumulation in the modern sector (Kuznets, 1969).

As this process unfolds, structural transformations of the economy are essential. The share of agriculture in both the gross national product and the labor force declines. In industry, capital-intensive production systems are established, based on an advanced pattern of division of labor and on a search for economies of scale—inevitably leading to increasing monopolization. As the labor force is reallocated from low-productivity agriculture to high- productivity industry, per capita incomes increase. Outmigration of labor and mechanization augment labor productivity in agriculture. In all sectors, increased proletarianization and higher labor productivity are reflected in increases in real money incomes to develop the home market (capacity to consume) that will sustain continued capital accumulation (capacity to produce). Structurally, the backward sectors of the economy decompose under dominance of the modern sector and are eventually incorporated. The dual economy becomes unimodal. The conjunction of increased proletarianization and real wages induces a demographic transition from high to low birth rates, and there is a tendency toward greater equality in the distribution of income (Kuznets, 1955).

The dominant stream of thought in this theory of economic development is based on a direct application of the western paradigm to today's less developed countries. Kuznets (1968) and Rostow (1963), in particular, derive theories of economic growth from detailed observation of the patterns of structural change during the agricultural and industrial revolutions of today's more developed countries. Similarly, models of growth in the dual economy constructed by Lewis (1958) and Jorgenson (1969) replicate the process by which surplus extraction from agriculture leads to growth of the modern sector and to dissolution of the dominated backward sector. These theories thus postulate the existence of a unique linear continuum in the process of economic growth, with some countries being more advanced than others on this one-way track. Marx (1970) made a similar induction from history by claiming that "the country that is more developed industrially only shows to the less developed the image of its own future." The role of agriculture in economic development and, consequently, the nature of agricultural development for the promotion of economic development have been established on the same basis by Mellor (1966), Johnston and Mellor (1961), and Nicholls (1969). The classical example used for this purpose is that of Japan (Ohkawa and Rosovsky, 1960; Ohkawa et al., 1970), the last country to become part of the center under the western paradigm.

The basic fallacy in this transposition of the western paradigm through time results from confusing fact and essence in the interpretation of history. Processes of historical change may be replicated following some general laws, but history itself cannot be repeated. A number of essential determinants that prevailed through the eighteenth and nineteenth centuries (when today's more developed countries—including the United States, Russia, and Japan—went through their economic takeoffs) simply do not exist anymore. Patterns of development followed previously cannot be repeated. To the contrary, transposition of the western paradigm into today's structural conditions of the periphery, largely characterized by the nature of the international division of labor, leads to the development of underdevelopment (Baran, 1960; Frank, 1969). Theories of economic development based on the transplanted western paradigm are thus either dishonest apologies or irresponsible fallacies. A new theory of economic development is clearly needed, tailored to today's structural conditions of the periphery. Since the 1950s some countries of the Third World have provided significant examples for establishing such a theory. (Reference is made to Nyerere's experiences with communitarian socialism in Tanzania and to other attempts based on the development of traditional rural cultures in Zambia, Guinea (Conakry), Congo (Brazaville), and Somalia as well as to the more advanced experiences in North Vietnam, North Korea, China, Cuba, Albania, and Burma.)

Nonreplicability of the Western Paradigm

The western paradigm has been successfully applied to two sets of countries during the nineteenth century: the industrial nations of Europe and Japan and the new territories colonized and settled by European immigrants (North America, South Africa, and Australia). Different structural conditions have permitted the rise of these old and new centers, but in both instances a number of these determining conditions no longer apply. The most important nonreplicable determinants of the takeoff are now discussed for the old centers.

ROLE OF COLONIALISM FOR PRIMITIVE CAPITAL ACCUMULATION. Pillage of the Third World, piracy, and brigandage (including the holy Crusades and the plunder of silver and gold in Latin America and India) permitted the old centers to develop under conditions of capital abundance. The resulting accumulation of merchant capital caused rapid inflation that led to a fall in real wages and created cheap labor. It also permitted merchants who traded luxury products with landowners to extract from them the proceeds of land rents. As a result, income was concentrated in the hands of the nascent bourgeoisie (Dobb, 1963; Jalée, 1968; Stein and Stein, 1970).

ABUNDANCE OF NATURAL RESOURCES PER CAPITA AND INTERNATIONAL MIGRATIONS. A population/land ratio much lower than in today's less developed countries allowed high productivity of agricultural labor with low capital intensity and hence a high agricultural surplus. Possibilities of horizontal expansion, especially through conquest and settlement of underpopulated areas, made production increases cheaper than through vertical expansion. The colonial countries were transformed into complements of the metropolitan economies to which they supplied cheap wage goods and raw materials. In addition, enormous international migrations toward the new territories contributed to increased productivity of labor in the old centers (Bagchi, 1972).

NATURAL AND IMPOSED PROTECTIONISM AGAINST MORE ADVANCED CENTERS. While free trade was imposed on the Third World by colonialism or co-option of the national elites *(compradores),* the spread of new centers occurred under effective protection against the competition of more advanced centers. For the first centers to emerge, the possibility of importing capital goods and manufactured products was nonexistent. Only after the second industrial revolution in the 1880s (with the cheapening of production of steel, the replacement of coal with oil as the essential source of power, and the development of mechanical engineering) did transportation costs collapse, and effective competition was made possible for raw materials and capital goods. Henceforth, low transportation costs permitted the establishment of a world division of labor, where the periphery exported plantation crops and mining products and the center exported industrial products.

In addition to natural protection, high tariff barriers have been the rule for the emerging centers. As Bairoch (1971) indicates, "throughout the 19th Century, no country has initiated its economic development without instituting high custom barriers to protect its nascent industries. . . . If this custom protection has been necessary at a time when the gaps in development were smaller than today and especially at a time when transportation costs created an important natural barrier, it is easy to understand how damaging is a lack of industrial protection for today's Third World."

TECHNOLOGICAL AND ENTREPRENEURIAL INDEPENDENCE AND SECTORAL ARTICULATION. National technological continuity was always preserved in the centers, as cottage industries were not destroyed by imposed free trade and capital goods were not imported. In addition the small disparity between modern and traditional technologies implied the need for small amounts of capital per worker, low levels of technical knowledge, small-scale decentralized factories, easy flows of financial and human resources from agriculture to industry, and forward and backward sectoral linkages not truncated by enclaves in either factor or product markets. These initial structural conditions of the industrialization process were lost after the second industrial revolution, thus leading to external dependency and sectoral disarticulation.

MARKET EXPANSION THROUGH EXTERNAL AND INTERNAL DEMAND FOR INDUSTRIAL PRODUCTS. Colonialism in the first half of the nineteenth century had for its principal objective the creation of external markets for industry through destruction by free trade of local manufactures in the highly populated trading regions (India and China) and by migration of European settlers into the new underpopulated territories. With external realization of the product ensured, wages in the center were maintained at a dramatically low level; this is the period of widespread pauperism and child labor. Later (in the 1890s), with the rise of monopoly capital in the center, the function of the periphery was transformed into providing an outlet for surplus capital seeking high rates of profit that would compensate for the declining rates in the center as well as into providing cheap raw materials and food for import. Both these functions were based on the exploitation of cheap labor in the periphery, thus contradicting the possibility of external market expansion. From then on (under the pressure of worker demands) market expansion obtained in the center by relating real wages to labor productivity and thus developing at par the capacities to accumulate and consume within the center economies. Social articulation (and the material basis for social democracies) was then established.

In the new territories of European settlement, other nonreplicable factors allowed for takeoffs—in particular, the enormous abundance of natural resources relative to population and the flows of physical and human capital transfers coming along with the immigrants.

These factors have been of such overwhelming importance that they explain the inability of today's less developed countries to industrialize along this model. Application of the western paradigm in the periphery creates a perverted process of accumulation that leads to underdevelopment.

Prescriptions of the Western Paradigm in the Periphery

Applied to the periphery, the western paradigm implies two general prescriptions. First, an accelerated industrialization to achieve import substitution. (e.g., see Chenery, 1955; Lewis, 1958). This is based on (1) large-scale, centralized, urban-based factories; (2) capital-intensive modern technology imported from the center; (3) export of raw materials and plantation crops to finance the import of capital goods, technological information, and often food; (4) call on foreign investment, foreign aid, and foreign entrepreneurs; and (5) realization of the product of the modern sector either in external markets (enclaves) or in internal markets mainly created by the return to capital and rents (consumption of part of the surplus value leading to low rates or savings).

Second, the agriculture sector contributes to this rapid industrialization through (1) generation of financial and labor surpluses by mechanization and capital-intensive technological change to increase the productivity of agricultural labor, (2) concentration of the land in large-scale commercial farming and plantations, and (3) extraction of surplus principally through deterioration of the terms of trade against agriculture ("cheap food policies," Schultz, 1968).

Structural Implications for the Periphery

Transposition of the western model to today's Third World has created structural conditions that differ from place to place; nevertheless, they display at least four common characteristics.

First, dependency on exports (plantation crops and raw materials); imports (capital goods; technology; and, increasingly, food); and foreign financing and entrepreneurs leads to a world division of labor (based on cheap labor in the periphery) that allows for deterioration of the terms of trade against the periphery (Amin, 1972; Emmanuel, 1972). The result is a decapitalization of the periphery and a structural deficit in the balance of payments that block the spread of industrialization.

Second, imported capital goods and technology sever backward linkages, while export of plantation and mining enclaves sever forward linkages as well. The result of this sectoral disarticulation is the lack of spread effects from development of the modern sector (including the modernization of agriculture) on capital formation, technological innovation, and learning by doing as well as on the implantation and modernization of other sectors of economic activities (Rosenberg, 1963). It also implies perpetuation and deepening of a deficit in the balance of payments that blocks the backward spread of the industrial structure and forces continued extroversion and dependency.

Third, realization of the product of the modern sector, either in external markets or in domestic demand created by the return to capital and rents, dissociates labor's income from market expansion. As a result the logic of market expansion to sustain accumulation through increased labor income (that characterizes the center) does not apply. Part of the surplus value is consumed to create the market, thus substantially reducing the rate of savings and implying the need to call on foreign capital and foreign aid. As labor costs are minimized following individual entrepreneurial logic of profit maximization, cheap food policies (in large part due to the dumping of food surpluses in the world market by the more developed countries) allow low monetary wages. Unfavorable prices for agricultural products (reflecting the lack of effective demand for food) in turn imply stagnation in domestic food production and create inflationary pressures, and the distribution of income becomes increasingly regressive.

Finally, traditional agriculture is preserved and expanded, as it serves to supply cheap semiproletarian labor for commercial agriculture (which faces low food prices and high industrial input prices) and the urban economy (where the objective logic of cheap labor under social disarticulation applies). The wage levels can thus fall below the subsistence needs of the worker and his family, as part of these needs are met through subsistence production in traditional agriculture. A sharp division of labor by sex and age becomes established, where women, the young, and the old become the principal agriculturalists. Population explosion is fueled by the conjunction of poverty and control of a plot of land, as children are essential means of production and protection. The most abject forms of poverty are found in this dominated sector.

OPTIONS IN THE DESIGN OF TECHNOLOGY FOR RURAL DEVELOPMENT

The particular goals and structural context of small-farmer rural development programs imply the need for a correspondingly particular technological basis. Identifying this basis is thus the necessary starting point from which guidelines can be set regarding the *design* of new technologies for rural development. Design is defined as a choice among activities; i.e., the designer must select both the products and the techniques of production. Criteria that enter into making this choice arise at two levels—at the level of the peasant farm and at that of the economy. The most important criteria at the level of the peasant farm are:

1. Production for home use versus production for sale—Will activities be selected to improve nutrition and health through production of subsistence crops for home consumption (high lysine corn, protein-rich vegetables, etc.) or to maximize an objective function expressed in monetary terms (production of commodities for sale in the marketplace, including feeds and cash crops)?
2. Market versus social prices—If activities are selected to maximize a monetary objective function, should commodities and factors of production be valued at current market prices or at social prices that represent society's valuation of alternative options?
3. Risk—What level of aversion to risk (due to weather, market, and institutional conditions) should be considered and under what specific form (safety first, discounting for risk)?
4. Dependency on purchased inputs—Should the technologies allow for maximum autarky from the market, or can purchased inputs be used (hybrid versus synthetic seed, commercial chemicals versus agronomic practices and farm-produced means of fertilization and controlling pests and weeds)?
5. Landsaving versus laborsaving activities—Should the new activities be landsaving (increasing yield and hence the productivity of both land and labor), laborsaving (increasing the productivity of labor by decreasing labor requirements without impact on yields), or both?

6. Physical capital depth—Can the new activities be capital intensive or should they, on the contrary, minimize investment requirements?
7. Divisibility—Should the new activities be fully divisible (seeds, biochemicals, and simple hand tools), or are indivisibilities acceptable (tube wells and mechanical equipment)?
8. Human capital depth—Can the new activities require sophisticated management techniques (integrated management approach, accounting techniques), or should they demand only minimal departure from traditional management patterns?
9. Employment effect—Can the new activities increase labor requirements (family labor and seasonal and permanent rural workers)?

The most important criteria at the level of the economy are:

1. Regional focus—Will new activities be developed for regions poorest in productive resources (eroded and hilly land, dry and variable climate, and limited infrastructure or, in particular, for irrigation and transportation) or for the best endowed areas?
2. Sectoral articulation—Can the new techniques require imported factors of production (either directly imported or produced with imported capital goods or technological know-how), or should they be within the capability of the existing industrial structure to create forward and backward linkages to support the modernization of agriculture (intermediate technology)?
3. Social articulation—Should the new activities be oriented toward the production of wage goods for the domestic market, thus providing the logic for wage increases to create an effective demand that will sustain accumulation in agriculture? (Since peasants mainly produce food that is the wage good par excellence, this criterion is of relatively minor importance here. By contrast, the two concepts of social and sectoral articulation would be the key criteria to development of planning models and a theory of choice of technique. Hirschman's (1958) theory of growth makes use of the criterion of sectoral articulation but fails to identify the essential concept of social articulation.)
4. Human capital formation—Should the new activities permit the establishment of working patterns that are nonalienating and therefore maximize the development of man's creative abilities (convivial tools, Illich, 1973)?
5. External effects and ecological conservation—Can we allow for externalities (chemical runoffs and straw burning) and accelerated resource depletion (forest, range, and soil conservation; use of fossil fuels)? Should the discount rates of 7 to 15 percent currently used in project

evaluation be accepted, or should the value of human time be revalued by using much lower rates? (With these discount rates implications of alternative projects are essentially undistinguishable beyond 10 to 15 years. On the other hand, society will still be here and must live with the effects.)

Designing new technologies for rural development implies making choices among these options. The trio of process, structure, and goal permits these choices to be made.

RURAL DEVELOPMENT IN THE WESTERNIZED PERIPHERY

In the dependent, disarticulated, and dualistic structure of the periphery under the western paradigm, three major types of rural development programs can be identified by their goals; each implies particular specifications for the design of the corresponding technological instruments. Identification of these programs can best be done by reference to a social map of the rural poor as in Figure 11.1.

In this chart income is decomposed into two sources: income from agricultural production (including the valuation of production for home consumption) and income from wage work and other sources. For empirical purposes the classification of households is related to CIDA's taxonomy of subfamily (SF) and family (F) farms and, among subfamily farms, of internal and external *minifundios* (Barraclough, 1973). Internal subfamily farms represent the payments in land privileges that semiproletarian workers of the

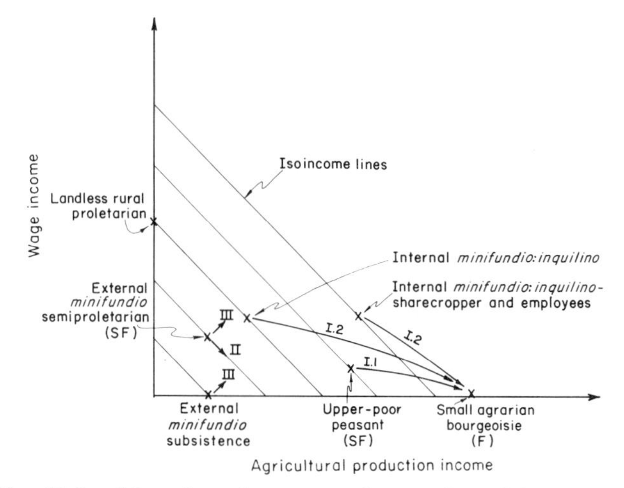

Fig. 11.1. Map of rural poverty and types of rural development programs.

Table 11.1. Rural development programs in the westernized periphery

Description	Type I	Type II	Type III
Goals	Transform upper SF farms into F farms	Reinforce functional dualism	Alleviate extreme poverty
	Increase agricultural surplus	Lower labor costs	None
	Create small rural bourgeoisie	Maintain status quo	Maintain status quo
Clientele	Upper external SF (I.1)	External semi-proletarian *minifundio*	External subsistence *minifundio*
	Internal SF (I.2)		
Instruments	Modernization (I.1)	Modernization	Modernization
	Land reform (I.2)	Institutional integration	Institutional integration
	Institutional incorporation		
Technological base			
Destination of production	Market (commodities)	Petty commodities and use values	Home use (use values)
Planning prices	Market	Market	Market
Risk	Medium risk	Low risk	Very low risk
Sources of inputs	Market	Market	High autarky
Technological basis	Land and labor-saving	Land and labor-saving	Landsaving
Capital depth	Intensive	Extensive	Very extensive
Divisibility	Divisible	Divisible	Divisible
Human capital	Medium	Minimal	Minimal
Employment effect	Minimal	Zero	High
Regional focus	Best endowed	Best endowed	Worst endowed

latifundio receive in compensation for labor services they render. External subfamily farms are usually semiproletarian, since some members of the household do wage work outside the farm, usually in the local commercial agricultural sector or in the mines; but other farms located in the most remote areas and in the closed corporate Indian communities may, exceptionally, be of pure agricultural subsistence (Wolf, 1965).

Three major types of rural development programs can be identified on this social map, each of which is characterized by a specific goal applied to a particular clientele (Table 11.1).

Type 1: Transform the Upper-Poor Peasants into Small Commercial Farmers

This type of rural development program has probably been most prevalent in Latin America. The clientele is composed of the external upper-poor peasants and the internal semiproletarians of the *latifundio*. For the first, the resource base is sufficient for a modernization program (diffusion of new technologies) to transform subfamily farms into small commercial farms (type I.1). For the second, land-reform programs that distribute land of the *latifundio* to its farmer workers and employees also provide a sufficient resource base to transform them into the category of small commercial farmers (type I.2). In both cases, social "incorporation" (Lehmann, 1971) provides the needed access to and limited control over the institutions (credit, information, and research) to sustain modernization. This type of program does not affect functional dualism and has no impact on the mass of the rural poor.

The economic goal of type I programs is to increase the marketable food surplus; the political goal is to create a stable, small, rural bourgeoisie that will identify with the interests of the medium and large commercial farmers. In general, the political goal will be dominant over the immediate economic goal.

The technological basis for type I programs is quite similar to that for modernization of commercial farming: focus on the areas best endowed ecologically and infrastructurally, production of commodities for sale in markets, medium risk and use of institutional arrangements to protect against risk (insurance and compensation funds), purchased inputs and use of institutional arrangements to facilitate access to these, landsaving and laborsaving activities, possibilities of deepening physical capital, medium management skills, and minimal employment effect.

Type II: Reinforce the Potential for Surplus Extraction by the Dominant Modern Sector through the Labor Market

The contribution of agriculture to economic development is essential in the western paradigm. With functional dualism, the contribution of subsistence agriculture is primarily via the labor market: wages paid to workers maintaining ties in traditional agriculture need only be a fraction of the subsistence needs of the worker and his family, the complement being insured by subsistence agricultural production. Increasing the productivity of labor in traditional agriculture allows labor costs for the modern sector to be reduced, thus increasing the rate of capital accumulation in that sector. (for the logic of this argument, see de Janvry, 1975).

Rural development Type II programs have reinforcement of functional dualism as a goal. The economic goal is to lower labor costs for the modern sector, while the political goal is to maintain the status quo of social marginality. Here the economic goal dominates the political one.

The clientele is the mass of external semiproletarian *minifundistas;* the instruments are modernization and institutional integration.

The technological basis of type II programs is defined by focus on the best endowed areas, production for both home consumption and sale, low risk, limited use of purchased inputs, landsaving and laborsaving activities, limited use of capital, minimal entrepreneurial talents, and zero employment effect.

Type III: Alleviate Extreme Rural Poverty

The clientele is composed of both external subsistence and semiproletarian *minifundistas*. The objective is purely political in terms of social status quo; the instruments include modernization and institutional integration. Due to the minimal land base, employment creation in nonagricultural activities will be a necessary complement to agricultural modernization. By contrast to type II projects, wages will have to be protected against downward pressures made possible by the rise of productivity of labor in agriculture for the welfare gains from the program to be retained by the "target population."

The technological base is defined by focus on the worst endowed regions, production for home consumption, very low risk, high autarky from the factors market, landsaving activities, minimal capital and managerial requirements, and strong employment effects.

In practice, specific rural development programs will be combinations of the above three pure types because their goals and clientele will overlap. Yet it is essential to clearly identify the type(s) to which particular programs relate to avoid gross mistakes in the specification of the required technological base. Attempts at transplanting the methodology of Plan Puebla, a type I program, to other areas of Latin America where the clientele and goals are of types II and III have resulted in spectacular failures. Lack of identification of project types thus results in the incapacity to learn from past experience and to design adequate new programs.

THE RURAL WAY TO ECONOMIC DEVELOPMENT

In the dependent, disarticulated, and dualistic structure rural development is only an appendix of agricultural and economic development. The production of a marketable agricultural surplus and exportables occurs in the commercial agricultural sector. The objective (rural poverty) and subjective (social tensions) contradictions created by this process of growth bring about the need for countervailing rural development programs. Thus, in this process of accumulation, agricultural development (commercial agriculture) is sought for the contributions agriculture can make to growth of the modern urban-industrial sector; and rural development (subsistence agriculture) is brought about by the need to counteract the contradictions created by accumulation in a dependent, disarticulated, and dualistic structure.

In the rural way to economic development, by contrast, rural development becomes the primary engine of growth. We shall refer to it as type IV, grass-roots rural development. Its prerequisite is deep structural change—hence drastic social options that eliminate dualism, restoring social and sectoral articulations at the national level and eliminating forms of dependency that are antagonistic to unimodal and articulated development. In so doing, type IV actually restores the fundamental structural characteristics that allowed for the economic takeoffs in the center, although on a radically transformed background in regard to international division of labor, technological capability, and human consciousness. But it can also supersede the growth experience of the center in attempting to avoid some of the major subjective contradictions of accumulation: alienation of labor, submission to consumerism, poverty in the midst of plenty, and environmental destruction.

Elimination of dualism requires a more egalitarian access to and control over productive resources. This implies restoring relations of "property" and "possession" at the community level. Social articulation is established by directing the production structure toward the demand for wage goods emanating from the mass of the population. In this way, creation of the home market to sustain accumulation is obtained by progressive increases in real incomes as the productivity of labor rises. As real incomes increase, the set of wage goods becomes redefined, following Engel's law, to eventually include durable goods in addition to food, textiles, and construction. Sectoral articulation is established by using capital goods and technological knowledge that are consistent with national productive capacities. In this fashion, modernization of agriculture has forward and backward spread effects that induce capital accumulation, technological innovation, learning by doing, and the implantation and modernization of other sectors of economic activity. The choice of a technology that enables the development of productive forces and is simultaneously consistent with the maintenance of both sectoral and social articulation is the correct definition of "intermediate technology." Production for the external market is no longer disassociated from production for the home market, as in the case of modern enclaves, but belongs to the same set of activities.

For type IV grass-roots rural development, the available resources are man with his learning and creative potential, land, and cottage industries. The scarce resources are capital and modern technological know-how. The strategy of development consists in modernizing a unimodal rural sector in an economic system that is articulated both socially (between capacities to produce and to consume) and sectorally (between sectors producing wage goods and capital goods). By decentralization, minimum division of labor, and active participation in decision making, the process of work aims at maximizing the formation of human capital.

The technological basis can be identified similarly to that for rural development in the westernized periphery: focus on all regions; production for community self-reliance and for the market; use of social prices in planning the use of resources for marketed commodities; low risk; high community autarky for the factors used and thus strong backward linkage effects of the modernization of agriculture; landsaving activities; extensive use of capital and maximum community self-financing; divisibility at the level of the economic units; strong employment effect; total technological consistency with the existing small-scale, decentralized, and locally controlled industrial structure (intermediate technology); production of wage goods for the domestic market; activities that allow for learning by doing and that minimize alienation; and minimal external effects and maximum ecological conservation.

CONCLUSIONS

Identification of the role of agricultural development in the process of growth allows us to contrast different types of rural development programs according to (1) the nature of the economic *process* where they apply; (2) the particular social, economic, and resource *structure* of the region; and (3) the economic and political *goals* sought by rural development programs. Major types of rural development programs in the dependent disarticulated, dual westernized periphery are (1) programs aimed at transforming the upper class of subfamily farms (both external and internal to the *latifundio*) into small commercial family farms; (2) programs aimed at increasing surplus extraction from traditional agriculture via the labor market to accelerate capital formation in the modern sector; (3) programs aimed at alleviating extreme rural poverty in a self-centered, articulated, unimodal structure; and (4) grass-roots rural development programs where agricultural modernization is the core of economic development. Each type of program has a corresponding need for a particular technological basis and particular institutional changes. It is thus essential to identify past and prospective programs with such broad types to correctly evaluate and/or design them.

The technological basis of rural development programs has been identified for a given process-structure-goal instance, i.e., for a given set of social relations of production. It is known, however, that technology (the development of the forces of production) is not neutral in its impact on the social relations of production but that there is a reciprocal relationship of determination between the two: development of the forces of production can lead to changes in social relations and to further development of production forces.

Technology thus appears as a historically powerful inducer of social change. When designing new technologies for rural development, the question then arises as to whether a further criterion should be taken into account, namely, the potential for alternative technologies to induce social changes of different types. The contrast between mechanical and biochemical innovations is a good example of options that have markedly different impacts on social change, with the former favoring concentration of land ownership, displacement of peasants, and hierarchical control of the labor force. But this key aspect of economic development remains largely unknown due to the complexity of the problem and the variety of possible instances, as there is clearly no mechanistic determination of production relationships by the level of development of production forces.

For rural development it is quite possible that the method of communication and diffusion of new technologies is more relevant for social change than the technological basis itself, especially when brought to choices among biological techniques as in the mandate of the international agricultural centers. Indeed, one of the most challenging aspects of rural development is in diffusing new technological options in such a fashion as to mobilize peasants to press toward social and structural changes that will ultimately liberate them from the process of poverty to which they are bound. However limited the scope of action of alternative technological designs, this fundamental criterion should always be remembered.

COMMENT / *Grant M. Scobie*

De Janvry has given us a perceptive, powerful, and appealing diagnosis of rural poverty. His attempt to review "the socioeconomic process through which rural poverty has been produced and is being reproduced" represents a level of intellectual articulation that is frequently lacking in discussions of this problem. As an economist raised on a steady diet of Friedman, I hesitate to comment on the validity of his broad-brush analysis. He has not aided *my* comprehension of why the scars of economic imperialism are less visible in New Zealand than in Uruguay; but I suspect this is much more my problem than his.

My principal concern is that he presents us with an astute diagnosis of the patient, clearly linking the observable symptoms to the cause of the disease; but then having diagnosed the problem as cancer, he proposes Band-Aids as the treatment! Only at one point, where he acknowledged "significant experiments" in Cuba, China, and elsewhere, was there hint of the deep surgery I felt sure he was going to prescribe, given the seriousness of the diagnosis. He tells us the world is tilted, toward the center and away from the periphery, and uses compelling historical arguments to support his diagnosis. However, I remain unconvinced that his prescription (which sounded strangely reminiscent of Mosher's *Getting Agriculture Moving* recast in the image of Berkeley), is adequate in the face of forces that produced the tilting in the first place. In his own words, productivity increases "for the rural poor could result in a subsidy to the commercial sector. Only by careful design of complementary *institutional and structural change* will welfare gains *possibly* be retained by the subsistence sector" (de Janvry, 1975). Perhaps I would feel happier with his analysis if I could see more solid evidence that in fact "technological change can then become a powerful means of inducing social change rather than an end in itself."

Are the rural poor and their miserable conditions a happenstance, somehow accidentally bypassed in the industrial, technocratic, urban-based boom of the postwar period? Or are they, as de Janvry poses, the product of a set of regional, national, and international forces that have evolved over centuries and continue to operate, favoring certain groups and condemning the rest to poverty? Can new agricultural technology really tilt this imbalance? Is it fair to even expect that technological change should be the vehicle for arresting and redirecting a set of political forces with two or three centuries of accumulated momentum? The question is *crucial* to our problem of establishing the criteria for deciding what type of technology to develop. It may be that burdening agricultural technology with the responsibility for correcting a broad spectrum of social ills is a hopelessly idealistic notion, which should not become part of the criteria for technology selection.

If rural poverty in the northeast of Bongoland is viewed as the result of dynamic social and political forces accumulated over centuries (and which continue operating to favor the rest of Bongoland at the expense of the northeast), can well-intentioned economists and scientists (backed with well-intentioned foreign funds) change this imbalance by choosing technologies that are labor using, risk neutral, and extensive in capital—i.e., truly adapted to illiterate, impoverished, capital-scarce, risk-petrified producers? I hope so, but the green revolution's contribution to date gives little ground for optimism. Perhaps we are designing technology for such a hostile environment (physically and, more important, sociopolitically) that a *prerequisite* is a change in that sociopolitical structure; without it, the potential recipients of the technology may continue to either never receive it or never capture its bounty.

De Janvry states that "deep structural changes" are a prerequisite for his (presumably preferred) type IV rural development. He then implicity rejects this deep surgical treatment in favor of technological Band-Aids; the latter, through some sparsely argued connection between the mode of production and technology, will (hopefully?) serve to awaken the social conscience of the peasantry, their resultant clamoring being silenced by greater social justice. If we are to rely on agricultural technology as the principal vehicle for eventually achieving a more equitable income distribution, I fear disarticulated dualism may be around for a long time.

But none of this should be taken in any way to diminish de Janvry's contribution to a horrendously complex issue. If his insights now stimulate him (or others) to formulate testable hypotheses about the catalytic social role of technology, our understanding will have been advanced appreciably.

In conclusion, let me briefly refer to the urban poor. I fear that in our enthusiasm for designing technology for the small-farm sector with the object of alleviating rural poverty, we tend to forget the increasingly numerous urban poor. Designing new technology for the inhospitable ecological zones (where rural poverty tends to be concentrated) may involve a substantial sacrifice in the potential contribution of a given quantity of research resources to total food output. The subsequent total supply of cheap food will be less. The risk I see is that we enter a zero-sum game, with poverty in the rural sector simply being substituted for poverty in the urban sector.

References

Albuquerque Lima, D. M., and Sanders, J. H. 1976. Seleção e avalição de nova. tecnologia para os pequenos agricultores do Sertão central do Ceará. Res. Ser. 16, Departamento de Economía Agrícola, Universidade Federal do Ceará, Brazil.

Almeida Carvalho, R. C., Almeida Duarte, P., Fernandes Pereira, P., and Fontelles Thomaz, A. C. 1976. Determinação das quantidades otimas de fertilizantes para as culturas de milho e feijao em municipios do estado do Ceará. Res. Ser. 7, Departamento de Economía Agrícola, Universidade Federal do Ceará, Brazil.

Amin, S. 1972. *Accumulation on a World Scale.* New York: Monthly Review Press.

Anderson, J. R. 1973. Sparse data, climatic variability and yielding uncertainty in response analysis. *Am. J. Agric. Econ.* 55:77-82.

______. 1974a. Simulation: Methodology and application in agricultural economics. *Rev. Mark. Agric. Econ.* 41:3-55.

______. 1974b. Risk efficiency in the interpretation of agricultural production research. *Rev. Mark. Agric. Econ.* 42:131-84.

______. 1975. Programming for efficient planning against nonnormal risk. *Aust. J. Agric. Econ.* 19:94-107.

______. 1976. Essential probabilistics in modelling. *Agric. Syst.*, pp. 219-31.

Anderson, J. R., Dillon, J. L., and Hardaker, J. B. 1977. *Agricultural Decision Analysis.* Ames: Iowa State Univ. Press.

Arrow, K. J. 1964. The role of securities in the optimal allocation of risk-bearing. *Rev. Econ. Stud.* 31:91-96.

Arrow, K. J., and Lind, R. C. 1970. Uncertainty and the evaluation of public investment decisions. *Am. Econ. Rev.* 60:364-78.

Badhuri, A. 1973. Agricultural backwardness under semifeudalism. *Econ. J.* 83:1685-1702.

Bagchi, A. K. 1972. Some international foundations of capitalistic growth and underdevelopment. *Econ. Polit. Weekly,* August, special number.

Bairoch, P. 1971. *Le Tier Monde dans l'Impasse.* Paris: Editions Gallimard.

Baran, P. 1960. *The Political Economy of Growth.* New York: Prometheus.

Barbosa, A. R. 1975. Relação de produção em differentes tamanhos de empresas região Serido—Rio Grande do Norte 1971-1972. Unpubl. M.Sc. thesis, Departamento de Economía Agrícola, Universidade Federal do Ceará, Brazil.

Barbosa, A. R., Sanders, J. H., and Abdon de Lyra, H. J. 1976. Opçoes tecnologicas para a região semi-arida do Rio Grande do Norte. Publição 04, Comissao Estadual de Planejamento Agrícola do Rio Grande do Norte, Brazil.

Barnard, C. S. 1963. Farm models, management objectives and the bounded planning environment. *J. Agric. Econ.* 15:525-49.

______. 1975. Data in agriculture. A review with special reference to farm management research, policy and advice in Britain. *J. Agric. Econ.* 26:289-333.

Barraclough, S. (ed.). 1973. *Agrarian Structure in Latin America.* Lexington, Mass.: Heath Lexington.

Benito, C. 1975. Cultural action and rural development. Rural Development Project Working Paper 5, Giannini Foundation Agric. Econ., Univ. Calif., Berkeley (mimeo).

Bieri, J., de Janvry, A., and Schmitz, A. 1972. Agricultural development and the distribution of welfare gains. *Am. J. Agric. Econ.* 54:801-9.

Binswanger, H. P., and Ruttan, V. W. (eds.). 1976. *Induced Innovation and Development.* Baltimore: Johns Hopkins Univ. Press.

Blackman, R. B., and Tukey, J. W. 1958. *The Measurement of Power Spectra.* New York: Dover.

Boon, G. K. 1964. *Economic Choice of Human and Physical Factors in Production.* Amsterdam: North-Holland.

Boussard, J. M. 1971. Time horizon, objective function, and uncertainty in a multiperiod model of farm growth. *Am. J. Agric. Econ.* 53:467-78.

Boussard, J. M., and Petit, M. 1967. Representation of farmers' behavior under uncertainty with a focus-loss constraint. *J. Farm Econ.* 49:869-80.

Camm, B. M. 1962. Risk in vegetable production on a fen farm. *Farm Economist* 10:89-98.

Campos Mesquita, T., Da Silva, P. R., and Sanders, J. H. 1976. Procura potencial para o sorgo granífero no Nordeste Brasileiro. Departamento de Economía Agrícola, Universidade Federal do Ceará, Brazil (mimeo).

Carter, H. O. 1963. Representative farms as guides for decision making. *J. Farm Econ.* 45:1448-59.

Charnes, A., and Cooper, W. W. 1959. Chance-constrained programming. *Manage. Sci.* 6:73-79.

______. 1963. Deterministic equivalents for optimizing and satisficing under chance constraints. *Oper. Res.* 11:18-39.

Chenery, H. 1955. The role of industrialization in development programs. *Am. Econ. Rev.* 45:40-57.

CIAT. 1974. *Annual Report, 1973.* CIAT, Cali, Colombia.

CIMMYT. 1974. *The Plan Puebla: Seven Years of Experience: 1967-1973.* Mexico City: CIMMYT.

Clayton, E. S. 1956. Research methodology and peasant agriculture. *Farm Economist* 8:27-33.

______. 1965. *Economic Planning in Peasant Agriculture.* University of London: Wye College.

Cocks, K. D. 1968. Discrete stochastic programming. *Manage. Sci.* 15:72-79.

Collinson, M. P. 1972. *Farm Management in Peasant Agriculture: A Handbook for Rural Development Planning in Africa.* New York: Praeger.

Dasgupta, D., Marglin, S. A., and Sen, K. 1972. *Guidelines for Project Evaluation.* New York: U.N. Ind. Dev. Org.

Day, R. H. 1965. Probability distributions of field crop yields. *J. Farm Econ.* 47:713-41.

Day, R. H., Aigner, D. J., and Smith, K. R. 1971. Safety margins and profit maximization in the theory of the firm. *J. Polit. Econ.* 79:1293-1301.

De Finetti, B. 1968. Probability: Interpretations. In *International Encyclopedia of the Social Sciences,* Vol. 12. New York: Macmillan.

De Janvry, A. 1975. The political economy of rural development in Latin America: An interpretation. *Am. J. Agric. Econ.* 57:490-99.

Dent, J. B., and Anderson, J. R. 1971. *Systems Analysis in Agricultural Management.* Sydney: Wiley.

Diamond, P. A., and Stiglitz, J. E. 1974. Increases in risk and risk aversion. *J. Econ. Theory* 8:337-60.

Dias de Hollanda, A., and Sanders, J. H. 1976. Avaliação da introdução de nova tecnología para pequenos e médios agricultores sob condiçoes de risco—o Serido do Rio Grande do Norte. Departamento de Economía Agrícola, Universidade Federal do Ceará, Brazil.

Dillon, J. L. 1968. *The Analysis of Response in Crop and Livestock Production.* Oxford: Pergamon Press.

Dobb, M. 1963. *Studies in the Development of Capitalism.* New York: International Publishers.

Doob, L. W. 1969. Testing theories concerning a subculture of peasantry. In *Subsistence Agriculture and Economic Development,* C. R. Wharton, ed. Chicago: Aldine.

Doyle, C. J. 1974. Productivity, technical change and the peasant producer: A profile of the African cultivator. *Food Res. Inst. Stud.* 13:61-76.

Duloy, J. H., and Norton, R. 1973. Chac: A programming model of Mexican agriculture. In *Multi-Level Planning: Case Studies in Mexico,* L. Goreux and A. S. Manne, eds. Amsterdam: North-Holland.

EMBRAPA. 1974. Pacotes tecnologicos para o algodao — Rio Grande do Norte. Circ. 11, Rio Grande do Norte, Brazil.

EMBRAPA et al. 1974. Diagnostico de Ceará. In *Alternativas de Desenvolvimento para Grupos de Baixa Renda na Agricultura Brasileira,* Vol. 2. Instituto de Pesquisa Economía, Universidade de São Paulo, Brazil.

Emmanuel, A. 1972. *Unequal Exchange.* New York: Monthly Review Press.

Encarnación, J. 1965. On decisions under uncertainty. *Econ. J.* 75:442-44.

Escobar P., G. 1973. Socio-economic diagnostic study. Cáqueza Project, ICA, Regional 1, Bogotá, Colombia (mimeo, Spanish).

Evenson, R. E., O'Toole, J. C., Herdt, R. W. Coffman, W. R., and Kauffman, H. E. 1977. Risk and uncertainty as factors in crop improvement research. Paper FF-1, International Rice Research Institute, Philippines.

Falcon, W. P. 1970. The green revolution: Generations of problems. *Am. J. Agric. Econ.* 52:689-710.

Fariz, M. A.,and Ferraz, L. 1974. Programa de sorgo e milheto, relatorio anual exercicio de 1973. IPA-PSM Bull. 2, Instituto de Pesquisa, Brazil.

Feder, E. 1967. The *latifundia* puzzle of Professor Schultz: Comment. *J. Farm Econ.* 49:507-10.

Fernández, F., and Franklin, D. L. 1973. Bean production systems. In Potentials of Field Beans and Other Legumes in Latin America. Series Seminars 2E, CIAT, Cali, Colombia.

Frank, A. G. 1969. *Latin America: Underdevelopment or Revolution.* New York: Monthly Review Press.

Freund, R. J. 1956. The introduction of risk into a programming model. *Econometrica* 24:253-63.

Gotsch, C. H. 1972. Technical change and the distribution of income in rural areas. *Am. J. Agric. Econ.* 54:326-42.

Grof, B., Paladines, O., Valdes, A., and Wells, E. 1975. Working Paper, Beef Cattle Production Program, CIAT, Cali, Colombia.

Hadley, G. 1964. *Non-Linear and Dynamic Programming.* Reading, Mass.: Addison-Wesley.

Halter, A. N., and Dean, G. W. 1971. *Decisions under Uncertainty with Research Applications.* Cincinnati: South-Western.

Hardaker, J. B. 1971. Farm planning by computer. Tech. Bull. 19, Ministry of Agriculture, Fisheries and Food, HMSO, London.

Hardaker, J. B., and Tanago, A. G. 1970. Assessment of the output of a stochastic decision model. *Aust. J. Agric. Econ.* 14:170-78.

Hayami, Y., and Ruttan, V. W. 1971a. *Agricultural Development: An International Perspective.* Baltimore: Johns Hopkins Univ. Press.

______. 1971b. *Resources, Technology, and Agricultural Development.* Baltimore: Johns Hopkins Univ. Press.

Hazell, P. B. R. 1970. Game theory: An extension of its application to farm planning under uncertainty. *J. Agric. Econ.* 21:239-52.

Hazell, P. B. R., 1971. A linear alternative to quadratic and semivariance programming for farm planning under uncertainty. *Am. J. Agric. Econ.* 53:53-62.

Hazell, P. B. R., and Scandizzo, P. L. 1973. An economic analysis of peasant agriculture under risk. Paper, 15th Int. Conf. Agric. Econ., São Paulo, Brazil, August.

Heyer, J. 1971. A linear programming analysis of constraints on peasant farms in Kenya. *Food Res. Inst. Stud.* 10:55-67.

______. 1972. An analysis of peasant farm production under conditions of uncertainty. *J. Agric. Econ.* 13:135-46.

Hirschman, A. O. 1958. *The Strategy of Economic Development.* New Haven, Conn.: Yale Univ. Press.

Hirshleifer, J. 1970. *Investment, Interest and Capital.* Englewood Cliffs, N.J.: Prentice-Hall.

IBRD. 1975. Rio Grande do Norte rural development project. Yellow cover report, Washington, D.C.

Illich, I. 1973. *Tools for Conviviality.* New York: Harper and Row.

Jackson Albuquerque, L., and Sanders, J. H. 1975. Rendimento do algodao herbaceo em função da fertilidade natural e artificial dos solos em alagoas. *Rev. Econ. Nordeste* 6:181-97.

Jalée, P. 1968. *The Pillage of the Third World.* New York: Modern Reader.

Jarvis, L. S. 1974. Cattle as capital goods and ranchers as portfolio managers: An application to the Argentine cattle sector. *J. Polit. Econ.* 82:489-521.

Jodha, N. S., and Ryan, J. G. 1975. ICRISAT study of traditional farming systems in the semi-arid tropics of India: Work plan and related aspects. Occasional Paper 9, Econ. Dept., ICRISAT, Hyderabad, India (mimeo).

Johnson, A. W. 1971. *Sharecroppers of the Sertão: Economics and Dependence on a Brazilian Plantation.* Stanford, Calif.: Stanford Univ. Press.

Johnson, R. W. M. 1969. The African village economy: An analytical model. *Farm Economist* 11:359-79.

Johnson, S. R., and Rausser, G. C. 1977. Systems analysis and simulation: A survey of applications in agricultural economics. In *A Survey of Agricultural Economics Literature,* Vol. 2, L. R. Martin, ed. Minneapolis: Univ. Minn. Press.

Johnston, B., and Mellor, J. 1961. The role of agriculture in economic development. *Am. Econ. Rev.* 51:566-93.

Jorgenson, D. W. 1969. The role of agriculture in economic development: Classical versus neoclassical models of growth. In *Subsistence Agriculture and Economic Development,* C. R. Wharton, ed. Chicago: Aldine.

Kataoka, S. 1963. A stochastic programming model. *Econometrica* 31:181-96.

Kirby, M. J. L. 1970. The current state of chance-constrained programming. In *Proceedings of the Princeton Symposium on Mathematical Programming,* H. W. Kuhn, ed. Princeton, N. J.: Princeton Univ. Press.

Knight, F. H. 1921. *Risk, Uncertainty and Profit.* Chicago: Univ. Chicago Press.

Kunreuther, H. 1974. Economic analysis of natural hazards: An ordered choice approach. In *Natural Hazard Perception and Choice,* G. F. White. London: Oxford Univ. Press.

Kunreuther, H., and Wright, G. 1974. Safety-first, gambling and the subsistence farmer. Univ. Pa. Fels Discussion Paper 59, Philadelphia.

Kuznets, S. 1955. Economic growth and income inequality. *Am. Econ. Rev.* 45:1-28.

______. 1968. *Toward a Theory of Economic Growth.* New York: Norton.

______. 1969. Economic growth and the contribution of agriculture: Notes on measurement. In *Agriculture in Economic Development,* C. K. Eicher and L. W. Witt, eds. New York: McGraw-Hill.

Lal, D. 1974. *Methods of Project Analysis: A Review.* Baltimore: Johns Hopkins Univ. Press.

Langham, M. R. 1968. A dynamic linear programming model for development planning. In *Economic Development of Tropical Agriculture,* W. W. McPherson, ed. Gainesville: Univ. Fla. Press.

Lehmann, D. 1971. Political incorporation versus political stability: The case of the Chilean agrarian reform, 1965-1970. *J. Dev. Stud.* 7:365-96.
Lewis, W. A. 1958. Economic development with unlimited supplies of labor. *Manchester School Econ. Soc. Stud.* 22:139-91.
Lin, W., Dean, G. W., and Moore, C. V. 1974. An empirical test of utility vs. profit maximization in agricultural production. *Am. J. Agric. Econ.* 56:497-508.
Lipton, M. 1968. The theory of the optimizing peasant. *J. Dev. Stud.* 4:327-51.
Little, I., and Mirrlees, J. 1969. *Manual of Industrial Project Analysis in Developing Countries,* Vol. 2. Paris: OECD.
Little, I., Scitovsky, T., and Scott, M. 1970. *Industry and Trade in Some Developing Countries: A Comparative Study.* London: Oxford Univ. Press.
Low, A. R. C. 1974. Decision taking under uncertainty: A linear programming model of peasant farmer behaviour. *J. Agric. Econ.* 15:311-21.
McFarquhar, A. M. M. 1961. Rational decision making and risk in farm planning: An application of quadratic programming in British arable farming. *J. Agric. Econ.* 14:552-63.
McInerney, J. P. 1967. Maximin programming: An approach to farm planning under uncertainty. *J. Agric. Econ.* 17:279-89.
———. 1969. Linear programming and game theory models: Some extensions. *J. Agric. Econ.* 20:269-78.
Magalhaes Bastos, J. A. 1973. Avaliação dos prejuizos causados pelo gorgulho em amostras de feijao de corda colhidas em Fortaleza, Ceará. *Pesquisa Agropecuaria Brasileira* 8:131-32.
Magalhaes Bastos, J. A., and Andrade Aguiar, P. A. 1971. Controle do gorgulho do feijao de corda com phostoxim. *Ciencias Agronómicas* 1:59-62.
Marx, K. 1970. *Capital.* Amsterdam: North-Holland.
Masson, R. T. 1972. The creation of risk aversion by imperfect capital markets. *Am. Econ. Rev.* 62:77-86.
———. 1974. Utility functions with jump discontinuities: Some evidence and implications from peasant agriculture. *Economic Inquiry* 12:559-66.
Mellor, J. W. 1966. *The Economics of Agricultural Development.* Ithaca, N.Y.: Cornell Univ. Press.
———. 1967. Towards a theory of agricultural development. In *Agricultural Development and Economic Growth,* H. M. Southworth and B. F. Johnston, eds. Ithaca, N.Y.: Cornell Univ. Press.
———. 1969. Production economics and the modernization of traditional agricultures. *Aust. J. Agric. Econ.* 13:25-34.
Miracle, M. P. 1968. Subsistence agriculture: Analytical problems and alternative concepts. *Am. J. Agric. Econ.* 50:292-310.
Mirrlees, J. A. 1974. Notes on welfare economics, information and uncertainty. In *Essays on Economic Behavior under Uncertainty,* M. Balch, D. McFadden, and W. Wu, eds. Amsterdam: North-Holland.
Moscardi, E. R. 1975. Allocative efficiency and safety-first rules among small-holding farmers. Rural Development Project, Working Paper 3, Univ. Calif., Berkeley.
Moseman, A. H. 1970. *Building Agricultural Research Systems in the Developing Nations.* New York: Agricultural Development Council.
Mosher, A. T. 1966. *Getting Agriculture Moving.* New York: Agricultural Development Council.
Naylor, T. H. (ed.). 1971. *Computer Simulation Experiments with Models of Economic Systems,* Ch. 9. New York: Wiley.
Nicholls, W. 1969. The place of agriculture in economic development. In *Agriculture in Economic Development,* C. K. Eicher and L. W. Witt, eds. New York: McGraw-Hill.
Nichols, D. A. 1974. The public discount rate. Dept. Econ., Univ. Wis., Madison (mimeo).
Obschatko, E., and de Janvry, A. 1972. Factores limitantes al cambio tecnológico en el sector agropecuario. *Desarrollo Económico* 11:263-85.

Ogunforwora, O. 1970. A linear programming analysis of income opportunities and optimal farm plans in peasant farming. *Bull. Rural Econ. Sociol.* 5:223-49.
Ohkawa, K., and Rosovsky, H. 1960. The role of agriculture in modern Japanese economic development. *Econ. Dev. Cult. Change* 9:43-67.
Ohkawa, K. et al. (eds.). 1970. *Agricultural and Economic Growth: Japan's Experience.* Princeton, N.J.: Princeton Univ. Press.
Owen, W. F. 1966. The double development squeeze on agriculture. *Am. Econ. Rev.* 56:43-70.
Patrick, G. F., and de Carvalho, J. J. 1975. Low income groups in Brazilian agriculture: A progress report. Exp. Sta. Bull. 79, Dept. Agric. Econ., Purdue Univ., Lafayette, Ind.
Pope, R. 1975. Diversification in farm production: A response to uncertainty? Dept. Econ., Univ. Calif., Berkeley.
Powell, J. D. 1972. On defining peasants and peasant society. *Peasant Stud. Newsl.* 1:94-99.
Pyle, D., and Turnovsky, S. 1970. Safety-first and expected utility maximization in mean-standard deviation portfolio analysis. *Rev. Econ. Stat.* 52:1083-86.
Rae, A. N. 1971a. An empirical application and evaluation of discrete stochastic programming in farm management. *Am. J. Agric. Econ.* 53:625-38.
______. 1971b. Stochastic programming, utility and sequential decision problems in farm management. *Am. J. Agric. Econ.* 53:448-60.
Rausser, G. C., and Johnson, S. R. 1975. On the limitations of simulation in model evaluation and decision analysis. *Simulation and Games.* 6:115-50.
Rivadeneira, H., Spain, J., Gutierrez, N., and Valdés, A. 1976. Metodología para el análisis de una pequeña finca ganadera: Ilustración en los llanos en Colombia. CIAT, Cali, Colombia.
Rogers, E. M. 1969. Motivations, values and attitudes of subsistence farmers: Towards a subculture of peasantry. In *Subsistence Agriculture and Economic Development,* C. R. Wharton, ed. Chicago: Aldine.
Rosenberg, W. 1963. Capital goods, technology, and economic growth. *Oxford Econ. Pap.* 15:217-27.
Rostow, W. W. 1963. *The Stages of Economic Growth.* Cambridge: Cambridge Univ. Press.
Roumasset, J. A. 1971. Risk and choice of technique for peasant agriculture: Safety first and rice production in the Philippines. Social Systems Research Institute Workshop Series, EDIE 7118, Univ. Wis., Madison.
______. 1974. Estimating the risk of alternative techniques: Nitrogenous fertilization of rice in the Philippines. *Rev. Mark. Agric. Econ.* 42:257-94.
______. 1976. *Rice and Risk: Decision-Making among Low Income Farmers.* Amsterdam: North-Holland.
Roy, A. D. 1952. Safety first and the holding of assets. *Econometrica* 20:431-48.
Ruttan, V. W. 1976. Induced technical and institutional change and the green revolution. In *Induced Innovation and Development,* H. P. Binswanger and V. W. Ruttan, eds. Baltimore: Johns Hopkins Univ. Press.
Ryan, J. G., and Perrin, R. K. 1974. Fertilizer response information and income gains: The case of potatoes in Peru. *Am. J. Agric. Econ.* 56:337-43.
Ryan, J. G., and Subrahmanyam, K. V. 1975. An appraisal of the package of practices approach in adoption of modern varieties. Occasional Paper 11, Econ. Dept., ICRISAT, Hyderabad, India (mimeo).
Sanders, J. H. 1973. Mechanization and employment in Brazilian agriculture, 1950-1971. Unpubl. Ph.D. diss., Univ. Minn., St. Paul.
______. 1976. Biased choice of technology in Brazilian agriculture: Mechanization. In *Induced Innovation and Development,* H. P. Binswanger and V. W. Ruttan, eds. Baltimore: Johns Hopkins Univ. Press.
Sanders, J. H., and Bein, F. L. 1976. Agricultural development on the Brazilian frontier: Southern Mato Grosso. *Econ. Dev. Cult. Change* 24:593-610.

Sanders, J. H., and Ruttan, V. W. 1976. Biased choice of technology in Brazilian agriculture. In *Induced Innovation and Development,* H. P. Binswanger and V. W. Ruttan, eds. Baltimore: Johns Hopkins Univ. Press.

Schluter, M. G. C. 1974. The interaction of credit and uncertainty in determining resource allocation and incomes on small farms, Surat District, India. Occasional Paper 68, Employment and Income Distribution Project, Dept. Agric. Econ., Cornell Univ., Ithaca, N.Y.

Schultz, T. W. 1964. *Transforming Traditional Agriculture.* New Haven: Yale Univ. Press.

______. 1968. *Economic Growth and Agriculture.* New York: McGraw-Hill.

______. 1975. The value of the ability to deal with disequilibria. *J. Econ. Lit.* 13:827-46.

Scobie, G. M., and Franklin, D. L. 1977. The impact of supervised credit programs on technological change in developing agriculture. *Aust. J. Agric. Econ.,*Vol. 21, No. 1, April.

Sen., A. K. 1959. The choice of agricultural techniques. *Econ. Dev. Cult. Change* 7:279-85.

______. 1966. Peasants and dualism with and without surplus labour. *J. Polit. Econ.* 74:425-50.

______. 1972. Control areas and accounting prices: An approach to economic evaluation. *Econ.J.* 82(No. 325s):1472-96.

Shipley, E., and Swanberg, K. 1974. The nutritional state of the rural family in eastern Cundinamarca. Cáqueza Project, ICA, Regional 1, Bogotá, Colombia (mimeo, Spanish).

Simon, H. A. 1966. Theories of decision-making in economics and behavioural science. In *Surveys of Economic Theory,* Am. Econ. Assoc., Vol. 3. New York: St. Martin's Press.

Spence, M., and Zeckhauser, R. 1972. The effect of the timing of consumption decision and the resolution of lotteries on the choice of lotteries. *Econometrica* 40:401-3.

Stavenhagen, R. 1969. Seven erroneous theses about Latin America. In *Latin American Radicalism,* I. L. Horowitz et al., eds. New York: Random House.

Stein, S., and Stein, B. 1970. *The Colonial Heritage of Latin America.* New York: Oxford Univ. Press.

Swanberg, K. G., and Escobar P., G. 1975. Labor use in eastern Cundinamarca. ICA, Bogotá, Colombia.

Szentes, T. 1971. *The Political Economy of Underdevelopment.* Budapest: Akademiai Kiado.

Thomson, K. J., and Hazell, P. B. R. 1972. Reliability of using the mean absolute deviation to derive efficient *E, V* farm plans. *Am. J. Agric. Econ.* 54:503-6.

Thorner, D., Kerblay, B., and Smith, R. E. F. (eds.). 1966. *A. V. Chayanov on the Theory of Peasant Economy.* Homewood, Ill.: Irwin.

Valdés, A. 1975. Some economic aspects of the cattle industry in Latin America. In seminar proceedings, Potential to Increase Beef Production in Tropical America, CIAT, Cali, Colombia.

Valdés, A., Gutiérrez, N., and Juri, P. 1977. Modelo de simulacíón por computador para fincas ganaderas. CIAT, Cali, Colombia.

Weinberg, G. M. 1975. *An Introduction to General Systems Thinking.* New York: Wiley.

Wharton, C. R. (ed.). 1969a. *Subsistence Agriculture and Economic Development.* Chicago: Aldine.

______. 1969b. Subsistence agriculture: Concepts and scope. In *Subsistence Agriculture and Economic Development,* C. R. Wharton, ed. Chicago: Aldine.

Wheeler, R. G. 1950. New England dairy farm management project as an example of the operating unit approach to farm management analysis. *J. Farm Econ.* 32:201-15.

Wills, I. R. 1972. Projections of effects of modern inputs on agricultural income and employment in a community development block, Uttar Pradesh, India. *Am. J. Agric. Econ.* 54:452-60.
Wolf, E. R. 1965. Types of Latin American peasantry: A preliminary discussion. *Am. Anthropol.* 57:452-71.
______. 1966. *Peasants.* Englewood Cliffs, N. J.: Prentice-Hall.
Zandstra, H. G., and Villamizar, C. A. 1974. Production investment plan for the small farmer. Cáqueza Project, ICA, Regional 1, Bogotá, Colombia (mimeo, Spanish).
Zeleny, M. 1975. Managers without management science. *Interfaces* 5:35-42.

Indexes

Author Index

Subject Index